Franz Mracek, Henry Weightman Stelwagon

Atlas of Diseases of the Skin

Including an Epitome of Pathology and Treatment

Franz Mracek, Henry Weightman Stelwagon

Atlas of Diseases of the Skin
Including an Epitome of Pathology and Treatment

ISBN/EAN: 9783337035198

Printed in Europe, USA, Canada, Australia, Japan

Cover: Foto ©berggeist007 / pixelio.de

More available books at **www.hansebooks.com**

ATLAS

OF

DISEASES OF THE SKIN

INCLUDING

AN EPITOME OF PATHOLOGY AND TREATMENT

BY

PROF. DR. FRANZ MRAČEK

of Vienna

AUTHORIZED TRANSLATION **FROM THE GERMAN**

EDITED BY

HENRY W. STELWAGON, M.D., PH.D.

Clinical Professor of Dermatology, Jefferson Medical College, Philadelphia;
Physician to the Department for Skin Diseases, Howard Hospital;
Dermatologist to the Philadelphia Hospital, etc.

With 62 Colored Plates and 39 Full-page Half-tone Illustrations

PHILADELPHIA AND LONDON

W. B. SAUNDERS & COMPANY

1900

EDITOR'S PREFACE.

THE importance of personal inspection of cases in the study of cutaneous diseases is readily recognized. For those lacking clinical facilities, the nearest approach to this method is the opportunity of having at command numerous well-executed colored plates; and an additional advantage is gained in having plates small enough for handy reference, and yet sufficiently large for satisfactory representation. Dr. Mraček has happily succeeded in supplying such atlas-pictures, and the selection of subjects portrayed is well adapted to the demands of average experience. While a few rare diseases—relatively rare at least in America—are presented, nearly all are pictures of cases of not infrequent occurrence.

The Editor has endeavored to follow closely the author's text, and in the translation of a part of the work he has had the aid of Dr. E. J. Stout, Instructor in Dermatology in the Jefferson Medical College. When deemed necessary, brief parenthetical notes have been added.

It is sincerely believed that this volume will be a material help to the working physician, especially in that most difficult branch of the subject—diagnosis.

PHILADELPHIA, 223 S. SEVENTEENTH ST.

7

PREFACE.

THE same motives which guided me in the preparation of my "Atlas of Syphilis" have also influenced me in preparing the subject-matter and in the selection of plates of the present volume—*i. e.*, the practical requirements and an appropriate choice of material. Circumstances have demanded certain limitations. For instance, of the acute exanthemata, only morbilli and varicella could be included, as the exanthematous infectious diseases are not allowed in my wards, but are sent to the hospital for infectious diseases.

I am indebted to the kindness of colleagues for some cases: to "Hofrat" Prof. Albert, Prof. de Amicis (Naples), "Primararzt" Dr. Eugen v. Bamberger, "Primararzt" Dr. Rudolf Frank, Prof. Lang, Prof. Kaposi, and "Primararzt" Dr. Ludwig Winternitz.

Plates 64 and 65 are taken from Kopp's Atlas. The water-colors and half-tone drawings have been executed by Mr. A. Schmitson, the artist, in his well-known and exemplary manner. The publisher has reproduced the work in the most careful manner.

The chapters on general therapeutics and treatment

of individual skin-diseases are from the pen of Dr. Siegfried Grosz, my assistant for many years.

I take this opportunity to acknowledge my thanks to all these gentlemen who have kindly aided me in my work. It is sincerely my wish that this book may meet with a kind reception and render practical assistance to those who consult its pages.

DR. MRAČEK.

VIENNA, November, 1898.

CONTENTS.

CONTENTS.

LIST OF PLATES.

15

DISEASES OF THE SKIN.

INTRODUCTION.

THE successful study of skin-diseases presupposes an exact anatomic knowledge of the skin, its appendages, physiologic functions, and reciprocal relations to other organs. Owing to limited space, we cannot devote a section to this subject; and this can indeed be studied more readily and satisfactorily in the larger works on dermatology, the handbooks on anatomy and physiology, etc., which treat of this subject. We would only emphasize especially that the skin belongs to the group of the most important organs of the body, and participates intimately in the functions of the entire organism; consequently, morbid changes in the skin may give rise to decided disturbances in the economy of the organism, and, *vice versâ*, diseases of internal organs may lead to pathologic changes in the skin. The general integument, therefore, should not be regarded as being solely intended for protection; nor looked upon, according to popular view, as an unimportant leathery cover.

The skin serves, it is true, as an organ of protection in the mechanical sense of the word; but it also possesses an important function in regulating the giving off of heat, and is itself an organ of secretion in which cutaneous respiration (the giving off of carbonic acid and water) also plays a *rôle*; although its power of absorption is very limited, it may under certain circumstances be not unimportant. Finally, the skin serves as an organ of touch, through which arise common sensation, the sense of locality or space.

The etiology of cutaneous diseases, as to be expected from the position of the integument in the organism, is many-sided and varied, and, owing to the relations existing between it and the organism as a whole, frequently complex. "External" causes of disease, mechanical and chemical, of the most diverse character, as well as parasites, which not infrequently gain access to the skin, may have a markedly damaging action. The skin is further affected by many noxious influences which attack the entire organism, be they infectious diseases, the most pronounced clinical group of which (acute exanthemata) have an especially active effect on the integument; or be they intoxication by poisons which have developed or have accumulated in the body itself, due to deficient elimination of the products of metabolism. The skin, therefore, presents a large number of *symptomatic* diseases : The acute cutaneous exanthemata, syphilis, equinia, typhoid fever, cholera, uremia, etc. all usually have cutaneous manifestations. Diseases of single organs (heart, liver, kidneys, nervous affections) are not uncommonly accompanied by phenomena on the external skin.

Direct injurious influences to which the skin may be subjected give rise to the so-called *idiopathic* skin-diseases. These influences, whether they are of an infectious or chemicophysical, traumatic character, are the causes of many of the acute and chronic inflammatory cutaneous diseases.

As cutaneous lesions are amenable to observation and to the sense of touch, they afford a very valuable subject—probably not as yet sufficiently appreciated—for theoretic and scientific study, which not infrequently can be aided and confirmed by microscopic examination. Although it would be very interesting to enter into a discussion relating to this subject, and to consider the sequelæ and development of granulation-tissue, formation of cicatrices, etc., we must, owing to lack of space, pass it by.

The symptoms of skin-diseases are divided into subjective and objective. To the former belong the various painful, itching, and burning sensations, etc.; further, those accompanied by a feeling of tension, disturbances of sensibility, anesthesia, and paresthesia. Of greater importance are the visible changes on the surface—objective symptoms—of the skin, the so-called cutaneous lesions or efflorescences, as they offer the necessary points for diagnosis.

We distinguish *maculæ*—macules, spots; of which the *erythemata* are examples. **In the same class** belong also *telangiectases, nævi vasculosi; hemorrhages,* as *petechiæ, vibices, ecchymoses;* and also *chloasmata, lentigines* or *freckles,* and *nævi pigmentosi,* all of which latter are pigmentary **spots**, usually of **a** brownish color. Yellow **plaques** or spots, usually somewhat thick **and** elevated, **are** called *xanthoma.* We have **further** *papules* and *tubercles,* lesions which **are** elevated **above** the level **of** the skin; tumors, circumscribed plastic elevations larger **than** a tubercle; and *wheals,* which **are** elevated above **the** surface; further, vesicular **and** bullous elevations **of** the epidermis with varying contents : *Vesiculæ,* **or** vesicles, *bullæ,* **or** blebs, *pustulæ,* **or** pustules, *ecthymata,* pustules of **larger** size accompanied **by** inflammatory infiltration **of the** surrounding parts. **Injuries to** the skin are **known as** *excoriations, rhagades* **(fissures).** *Ulcera,* ulcers **of the** skin, **are** the **result of more** severe inflammatory processes accompanied **by necrosis.** *Squamæ,* or scales, are the products of morbid desquamation or exfoliation **of the** epidermis. When the **contents of** the pustules or **ulcers** dry up, **crusts,** *scabs,* or *scurf* are formed, which cover the diseased **or** injured places.

Efflorescences may **be** solitary (*efflorescentiæ solitariæ*) **or** scattered (*dispersæ*), crowded together (*aggregatæ*), or in circles **or** segments of circles (*efflorescentiæ annulares, circinatæ*). The latter frequently form when the process spreads at the periphery **and** undergoes involution in the center. The name *iris* **(herpes, or** erythema iris) is **given** to that form in **which** several circles of efflorescences occur around a primary focus. *Gyri* are more or **less** circular lines, which **are** formed by the confluence of several circles and segments of circles. *Exanthem* designates a cutaneous eruption which is distributed over large surfaces or over the entire body.

There remains to be mentioned that certain skin-diseases have *seats of predilection*, which are partly dependent upon external causes (pressure of clothing); partly upon anatomic conditions—ramifications of nerves, vascular areas; and, finally, upon conditions of the texture of the skin itself—*e. g.*, the so-called lines of cleavage of the skin as described by Voigt and Langer.

We cannot refrain from making a short reference to the classification of skin-diseases. Every system serves as a didactic aid, and the endeavor to establish such in our eminently empirical science appears natural. The rich abundance of etiologic facts which the external causes of disease afford us has tended, with the inclusion of those found due to bacteriologic factors discovered during the last twenty years, to make an etiologic system possible. We are, on the other hand, not yet united, in many groups of diseases, as to the anatomic details, still less on etiology, which frequently has not passed beyond the hypothetic stage. What a *rôle* is still played at the present day by trophic disturbances in the nerve-tracts and by the so-called reflex neuroses!

We do not consider it our duty, however, in this elementary treatise, to enter at length on such far-reaching questions as this of classification; we also believe that Hebra's system, as slightly modified in many recent works, is still sufficient for the study of skin-diseases, even at the present day.

GENERAL THERAPEUTICS.

In dermatology we are as yet far from attaining the desired object of our therapeutic endeavors—to treat forms of disease according to their etiology. In but few skin-diseases has the question relating to the primary or essential causative factor or factors of an affection been satisfactorily settled ; the treatment of the greater number of diseases is still based upon the symptoms. As these are subject to changes during the course of a malady, the therapeutic indications also change ; it is accordingly of great importance to the physician to recognize each one of these several phases or stages, and he will then be enabled to use intelligently the numerous remedies which dermatologic therapeutics places at his disposal.

Internal Treatment.—The older practitioners recommended and employed various internal remedies in skin-diseases. Nearly all of these have been forgotten at the present day ; the ideas, however, of treating skin-affections by placing stress upon dietary rules and internal medication are again coming more and more to the front. In some skin-diseases dietary regulations are not only strongly to be advised, but are even indispensable. One need, for instance, only recall urticaria, often due to ingestion of certain kinds of food, and which may appear at other times when intestinal digestion is imperfect or faulty ; and the erythemata, which occur under similar etiologic conditions ; and also the eruptions of eczema in diabetic, nephritic, and gouty individuals.

It is therefore surely an error to practise dermatology with the aid of the ointment-pot alone ; just as it would be, on the other hand, to endeavor to combat marked changes in the cutaneous integument by simply forbidding certain articles of food.

Of the internal remedies we desire to mention the following : Arsenic, mercury, iodin, carbolic acid, tar-preparations, pilocarpin, atropin, quinin, sodium salicylate, thyroid preparations, calcium chlorid, menthol, etc. Some cases are benefited by a course of treatment with the natural mineral waters (Carlsbad, Franzensbad, Roncegno, Hall, Lipik [and in our own country Richfield Springs, Hot Springs of Virginia, Healing Springs, Bedford Springs, and many others well known.—ED.]).

External Treatment of Skin-diseases.—We employ the following means :

1. *Water* for ablutions **and for** partial **and** full baths ; the cold water **as** an astringent and warm water dilating the vessels. The continuous water-bath (Hebra's water-bed) may be employed in pemphigus foliaceus, decubitus, extensive burns, universal psoriasis, lichen ruber, and pityriasis rubra Hebra. Medicated baths **are** baths containing alkalies, potassium sulphid, brine, **tar,** corrosive sublimate, etc.

2. Besides **waters, the** *fats* serve materially for the purpose of softening morbid accumulations on the skin. Mineral, vegetable, **and** animal fats, in solid and fluid form, are employed ; they **are used** alone or as vehicles for **medicaments.** We mention those **most** frequently used : Petrolatum, vasogene (oxygenated hydrocarbons), vasole (Hell), cacao-butter, ol. olivæ, ol. amygdalarum dulc., ol. lini, **ol.** rapæ (rape-seed oil), ol. ricini, **lard,** spermaceti, œsypus (the natural fat **of wool),** adeps lanæ (wool-fat), lanolin [this **is** not a fat, but consists principally **of** ethereal fatty acids of cholesterin and **isocholes-**terin, which **are** also found in all **tissues** containing **kera-**tin and **in** the human skin and human hairs, and of **free** fatty acids (up to 30 per cent.)], cod-liver oil, and oleum physeteris.

Further : Mollin (an over-fatty **soap** made **of** pure kidney-fat and the finest Cochin cocoanut-oil, saponified by mixing potash- and soda-lye and addition of glycerin), myronin (produced from vegetable **wax** and ol. physeteris), **resorbin (made of** almond-oil, wax, and addition of gelatin, soap, and adeps lanæ), glycerinum saponatum (H. **v.** Hebra), epidermin (Kohn), unguentum lanolini Paschkis (lanolin. anhydr., **65** (ʒix gr. xxxv) ; paraffin. liquid., 30 (ʒviiss) ; ceresini (mineral wax), 5 (gr. lxxv) ; aq. destill., 30 (ʒviiss)), vaselinum lanolinatum (Hell), and glycerin.

Ointment-mulls (Salbenmulle) (Beiersdorf) are bandages spread with special salve-mass variously medicated, which do not adhere **to** the skin, but must be kept in position **by** suitable bandages. The fatty mass of the ointment-mulls consists principally of sebum benzoinatum, with addition of more or less wax.

3. *Soaps.*—These are combinations of fatty acids with alkalies. We distinguish between soft soaps (fat saponified by potash-lye) and hard soaps (fat saponified by soda-lye). When all of the alkali of the soap is combined with fatty acids the soap is *neutral.* The action of the soaps is said to be due to the soluble basic, fatty-acid salt. Over-fatted soaps are those which contain along with the fatty-acid salts, of which neutral soap consists, a certain quantity of unsaponified fat. Unna's over-fatted or basis-soap is made of the best beef-tallow and a mixture of two parts of soda-lye and one part of potash-lye; sufficient olive-oil is added to the soap-mass so that about 4 per cent. will remain unsaponified. Eichhoff has produced soaps containing various pulverulent substances.

We employ, in addition, Hebra's "spiritus saponatus kalinus" (tinctura saponis viridis) according to the following directions :

℞ Saponis viridis,   200 (ʒl gr. xv).
    Solve leni calore in
  Spirit. vini,   100 (f ʒxxv gr. viiss).
    Filtra et adde
  Olei lavandulæ,
  Olei bergamottæ,   āā 3 (gtt. xlv).
Misce et filtra.
Sig.—Spiritus saponatus kalinus (tinctura saponis viridis).

Finally, a number of medicated soaps (naphtol-sulphur soap, sulphur and tar, corrosive sublimate, menthol, thymol, resorcin, etc.).

4. *Varnishes.*—Excipients which, when painted on the skin, dry and form a smooth coating.

a. Varnishes soluble in water : Linimentum exsiccans Pick consists of tragacanth, 5 parts ; glycerin, 2 parts ; distilled water, 100 parts.

Unguentum caseini Unna consists of alkali-casein,

glycerin, vaselin, and water. It is miscible with all substances which do not coagulate casein. Tar up to 20 per
cent. may be added to the casein-ointment, although with
this an addition of 1 part of sapo viridis to 4 parts of
water is recommended, so as to render the product less acid.
Rubbed on the skin, it dries into an elastic, smooth layer.

Gelanthum consists substantially of tragacanth, gelatin,
and water.

Gelatin-paste, according to Unna's formula, is as
follows :

R, Gelatinæ alb.,   30 (ʒviiss) ;
Zinci oxid.,   30 (ʒviiss) ;
Glycerini,   50 (fʒxiiss) ;
Aquæ,   90 (fʒxxiiss).

The gelatin is dissolved in the water over a water-bath,
the glycerin added, and the zinc oxid well incorporated.
When desiring to use, melt over water-bath and paint on
with brush.

b. **Varnishes** insoluble in **water** : Collodion, traumaticin
(liquor gutta-perchæ), liquor adhæsivus Schiff or filmogen
(cellulose nitrate dissolved in acetone with addition of oil).

5. *Pastes.*—These are mixtures of medicaments having
the consistence of dough.

6. *Plasters.*—These consist of lead and soap, or of a
mixture of turpentine, various resins, and fats, or of
varying proportions of the two plasters.

The Unna-Beiersdorf gutta-percha plaster-mulls are
plasters in which the fabric is first coated with a thin
layer of gutta-percha. The thickly applied plaster-mass
consists principally of caoutchouc with addition of adeps
lanæ, and is variously medicated. The "paraplasters"
have as base a close cotton material of very fine fiber,
which is saturated with a solution of caoutchouc and vul-
canized. Collemplastra are plasters in which caoutchouc
is mixed with the plaster-mass.

7. *Powders.*—Starch, talcum, magnesium carbonate,
and zinc oxid are most usually employed.

DISORDERS OF THE GLANDS.

The secretory processes in the skin consist, in the main, of the ordinary secretions of the sweat- and sebaceous glands. Various substances found in the circulation are mixed with these secretions, so that they always represent a complex mixture. The normal secretions of the sudoriparous glands contain fat and the products of the so-called materia perspiratoria. This latter comes from the blood-vessels and mixes with the sweat, and usually consists of volatile fatty acids, which are mixed with the glandular secretions, and which may be quite abundant and may rapidly undergo change and give off a specific odor (**osmidrosis, bromidrosis**). The sweat-secretion is most abundant in the axillæ and in the genital region, which are rich in glands and which in certain individuals gives rise to an especially pungent, penetrating odor. It seems surprising that even pus-cocci have been excreted in perspiration of the skin, especially when sweating is profuse (Brunner, Eiselsberg).

The vicarious function of the sweat-glands, between which and the renal secretion there exists a relationship, is of especial importance. We observe under certain physiologic as well as pathologic conditions, when the function of the sweat-glands is increased, that the usual daily quantity of urine is decreased. We can, furthermore, frequently demonstrate admixtures of urea in the sweat in diseases of the kidneys, and also excretions of balsamic remedies, etc. The skin of diabetics, who pass large quantities of urine, *vice versâ*, is characterized by dryness.

Pathologic increase of sweat-secretion (**hyperidrosis**) is usually observed in corpulent individuals and those who undergo but slight bodily exertion, in psychic excitement, and also after conditions which lead to hyperemia of the skin. Profuse sweating often occurs in cachectic, tubercular, and anemic subjects. Subjective symptoms of prickling and slight itching of the skin sometimes may precede the sweating.

Increase in sweat-secretion of certain regions of the body, as the palms of the hands and the soles of the feet (*hyperidrosis palmarum et plantarum*), is to the individual thus afflicted of considerable importance. It is common in anemic subjects, whose hands and feet are cyanotic, owing to stasis, and who complain of sensations of cold in the extremities. This excessive sweating may exist for many years without any change whatever taking place in the skin. In rare instances vesicles sometimes may form on the fingers, more frequently on the toes; these rupture and lead to excoriations of the epidermis [dysidrosis, pompholyx ?—Ed.]. The epidermis between the toes is frequently macerated and peels off; painful excoriations and fissures occur, which may give rise to troublesome inflammation, and exceptionally to the formation of pus.

Dysidrosis, pompholyx, or **cheiropompholyx** [Hutchinson] occurs on the palms of the hands, on the sides of the fingers, and on the soles of the feet, owing, it has been believed, to retention of sweat. Vesicles and blebs, from the size of a pin's head to that of a pea, or larger, develop; their contents are perfectly clear at first, though they become turbid later on. The inflammatory symptoms, redness and slight or marked swelling of the epidermis, complete the picture of this disease. The affection disappears after the vesicles have ruptured spontaneously or have been ruptured by macerating treatment or accidentally. As the disease, however, relatively often attacks individuals who suffer from sweating feet, its recurrence is not uncommon (Plate 1).

Treatment.—In universal as well as local hyperidrosis it is of great importance to consider the possible underlying cause or causes (tuberculosis, anemia, etc.). Of internal remedies which have the power of influencing excessive secretion of sweat, we mention especially atropin and agaricin.

R҂ Atropin. sulphat., 0.015 (gr. $\frac{9}{40}$) ;
  Extr. taraxaci,
  Pulv. rad. althææ, q. s.
Ft. pil. No. xx.
Sig.—One pill night and morning.

Or

R҂ Atropin. sulphat., 0.01 (gr. $\frac{3}{20}$) ;
  Aq. menth. pip., 10 (f ℥iiss).—**M.**
Sig.—Five to ten drops **t. d.**

R҂ Pulv. agarici alb., 1 (gr. **xv**).
Dtur. tal. dos. No. **x.**
Sig.—One powder **t. d.**

R҂ Agaricini, 0.015 (gr. $\frac{9}{40}$).
In pil. No. xxx.
Consperg. sem. lycopod.
Sig.—One pill **t. d.**

The following are advised in the external treatment :
Baths, ablutions, and applications of alcoholic solutions,
such as menthol (1 : 100), carbolic acid (1 : 100), salicylic
acid (1–2 : 100), naphthol (β-naphtoli, 1 (**gr. xv**) ; aqua
coloniensis, 25 (ʒvj gtt. xv) ; spir. vini **gall.**, 175 (℥vss)).
A dusting-powder should be subsequently applied. The
following is useful for this purpose :

R҂ Salol., 1 (gr. **xv**) ;
  Zinci oxidi,
  Talc. ven., $\bar{a}\bar{a}$ 45 (℥iss).—**M.**
Sig.—Dusting-powder.

Or

R҂ Acidi salicyl., 5 (gr. **lxxv**) ;
  Acidi tartar.,
  Acidi boric., $\bar{a}\bar{a}$ 10 (ʒiiss) ;
  Zinci oxidi, 25 (ʒvj gr. xv) ;
  Talc. venet., 50 (ʒxiiss).—**M.**
Ft. pulv.
Sig.—Dusting-powder (Eichhoff).

In hyperidrosis pedum Hebra's favorite treatment with unguentum diachyli is often useful. The feet are daily enveloped with bandages spread with ung. diachyli, pledgets of lint smeared with this ointment being placed between the toes. This proceeding is continued for ten to fourteen days, during which period the feet are not to be washed. A few days after the dressing has been discontinued the skin exfoliates, and when desquamation has ceased the hyperidrosis is usually noted to have been relieved.

Applications of a 5 per cent. solution of chromic acid, solutions of formalin and corrosive sublimate are to be recommended; also painting with the following:

R̥. Liq. ferr. sesquichlorati, 30 (f℥viiss);
        Glycerini,   10 (f℥iiss);
        Ol. bergamottæ,   20 (f℥v).—M.

Sig.—To be applied with a brush to the sole of the foot and the regions between the toes (Legoux).

The sebaceous secretion of the skin is the product of the sebaceous glands, whose fat-cells secrete the nascent sebum found on the surface. An abnormal increase in the amount of sebaceous matter is known as seborrhea, which, when it appears in the form of an oil-coating, constitutes the condition known as *seborrhœa oleosa;* when the excessive sebaceous secretion dries up with the loose epidermic cells into scales, it gives rise to the type known as *seborrhœa sicca seu squamosa.*

Oily seborrhea may exist for years on the nose, forehead, and chin of many individuals without demonstrable cause; it may also be seated upon the scalp. Seborrhœa sicca is observed most frequently. It may be observed at almost any age, but is more common during adolescence and early adult life. It is also noted on the scalp of nursing-infants as a dry, hard crust, which adheres to the tender hairs.

The *vernix caseosa* is of similar origin, and occurs in newly-born infants as smegma, covering the whole body and consisting chiefly of detached epithelium.

The same disease is exemplified in collections of smegma in the preputial pouch in balanitis and balanoposthitis, and on the prepuce of the clitoris and inter-labial folds; these conditions lead to maceration of the epidermis and to excoriations, and even to inflammation accompanied by secretion of pus.

When seborrhea has existed for a longer period it gives rise to **comedones.** These formations are also noted when there is but a slight oilly or branny seborrheic condition of the surface of the skin. The fat and loose epithelium become inspissated in the excretory duct, lanugo-hairs and the Demodex folliculorum (Plate 64, Fig. 1) are mixed with this secretion, and the dilated follicle is filled with a greasy mass having a black external covering. These plugs are frequently loosened by the accumulating secretion beneath in the follicle, and can be readily removed. The excretory duct, which has become patulous, can be seen as an opening in the skin. Owing to increased accumulation of sebum in the cystic, dilated excretory ducts, the comedones may be converted into adenomata from the size of a pea to that of a bean (*vide* Plate 3).

As a consequence of comedo or blocking of the sebaceous ducts, inflammation of the sebaceous glands—*acne*—finally results, which will be discussed later on.

Treatment of Seborrhea.—The accumulated scales and crusts should be softened with oils or fats and then removed. When this has been done, or to aid in this, the scalp is thoroughly washed with soap (tinctura saponis viridis) and lukewarm water. The scalp, which may have become sensitive and moist, is covered with ointment. Zinc oxid, sulphur and salicylic acid, sulphur and zinc oxid, in ointment-form, and pastes of sulphur and zinc oxid are employed:

R̟ Zinci oxidi, 6 (ʒiss);
Sulph. præcip., 4 (ʒj);
Terr. siliceæ, 2 (ʒss);
Adipis benz., 20 (ʒvij).—M.
Ft. pasta (Unna).

Ointments of white and red precipitate, 5 to 30 grains to the ounce, are preferable if the hair is long or has not been cut. [Ointments containing pulverulent substances in any quantity are not so well adapted for scalp treatment as those just mentioned or those containing salicylic acid, resorcin, or sulphur, 5 to 30 grains to the ounce.—ED.]

Conjointly or alternately with ointment we use ablutions containing spirituous solutions of carbolic acid (0.35–0.70 (gr. v–gr. x) to the ounce), salicylic acid, β-naphtol, and resorcin ; the last in ointment, 2 to 10 per cent. strength, or either in alcoholic or aqueous solution of 2 to 4 per cent.

When the disease is localized on other parts of the body treatment based on the same principles is employed, but the applications should be weaker.

PITYRIASIS CAPITIS (SEBORRHŒA SICCA).

[The author, while placing this under seborrhea, recognizes its clinical difference by giving it a special heading for treatment. Most writers consider this as belonging to Unna's seborrheic eczema.—ED.]

The method of treatment, as recommended by Lassar, should be mentioned first. This consists of :

1. Shampooing with tar-soap for ten to fifteen minutes ; this is washed off with warm water, which should be gradually cooled.

2. Washing with

R, Sol. hydrarg. chlorid. corros., $\left\{ \begin{array}{l} 0.50 : 150 \ (\text{gr. viiss} \\ \text{to ʒiiss water}) ; \end{array} \right.$

Glycerini,
Spir. coloniensis,	āā 50 (fʒxiiss).

3. Shampooing with
R, β-naphtol,	0.25 (gr. iv) ;
Alcohol. absolut.,	200 (fʒvj ʒij).

4. Rubbing into scalp
R, Acidi salicylici,	2 (gr. xxx) ;
Ol. olivæ,	ad 100 (fʒxxv).

In connection with soap-washing and spirituous applications to the scalp, sulphur-ointments will also give good
results in these cases. Unna·recommends :

> ℞ Adipis lanæ,
> Aq. calcis,
> Aq. chamomillæ,
> Ung. zinci oxidi,   āā 10 (ʒiiss);
> Sulphur. præcip.,   2 (gr. xxx);
> Pyrogalloli oxid.,   0.40 (gr. vj).

The following is also useful, to be gently rubbed in :

> ℞ Tinct. cantharid.,   10 (fʒiiss) ;
> Tinct. benzoini,   20 (fʒv) ;
> Hydrarg. chlorid. corros., 0.20 (gr. iij) ;
> Chloral. hydrat.,   4 (ʒj) ;
> Resorcini,   5 (gr. lxxv) ;
> Ol. ricini,   10 (fʒiiss) ;
> Alcohol. absolut.,   200 (fʒvj ʒij)—M.
> Sig.—For local use.

Recently captol (a product of condensation of tannin
with chloral) has been recommended by Eichhoff :

> ℞ Captoli,
> Chloral. hydrat.,
> Acid. tartar.,   āā 1 (gr. xv) ;
> Ol. ricini,   50 (fʒxiiss) ;
> Spirit. vini (65 per cent.),   100 (fʒxxv).—M.
> Sig.—For external use.

MILIUM (PLATE 2).

In this condition round grains the size of a millet-seed,
of a milky-white color, and slightly raised above the level
of the skin, can be seen shining through the epidermis.
They are met with chiefly on the eyelids, cheeks, temporal regions, and male genitalia ; rarely on the labia
minora. When the epidermis is incised and these small
bodies have been removed from their bed, they fall to

pieces on slight pressure. They consist of dry epidermic cells and fat.

Treatment.—The overlying skin is incised with a small knife and the contents removed by lateral pressure. The ensuing wound, which is insignificant, heals very rapidly. When a large number of small milia exist a desirable method of treatment is that which produces exfoliation of the epidermis; and this may be attained by exciting a moderate degree of inflammation by stimulating the skin with applications of soft soap (Kaposi).

The names *Molluscum Contagiosum, Molluscum Verrucosum, Molluscum Epitheliale,* are applied to a verrucous proliferation on the skin, appearing as a rounded, shining, pearly, translucent, slightly elevated growth, and usually attaining the size of peas, which project hemispherically and show a slight depression at their apex. Lateral pressure with the fingers or curetting causes the contained whitish mass to be ejected, which is seen to be lobular in construction and surrounded by a thin covering of connective tissue; this sends out processes which converge toward the center as septa. The mass often has a firmer cover; it can be easily crushed to pieces, and the contents are found to be made up of epidermic cells, fat, crystals of fat, and so-called molluscum bodies. These latter are structureless, slightly shiny formations of ovoid shape, smaller than an epithelial cell, and are usually surrounded by epithelial cells and cell-débris (Plate 65, *b*).

Molluscum contagiosum has been demonstrated to be contagious; the growths are often found on contiguous surfaces of the skin and in individuals who are in close contact with one another (children and nurses). The most common sites are the face, eyelids, the genitalia, scrotum and penis, the external female labia (see *Atlas of Syphilis,* Plate 71), and inner folds of the thighs. They also occur on the neck, hands, and forearms, and may even be distributed over the general surface, as observed by Kaposi in small children.

Treatment.—The contents are usually removed by lateral pressure ; when the lesions are numerous or persistent, removal by surgical means (Volkmann's spoon ; excision) is recommended. Puncturing with a pointed knife, pressing out the contents, and touching the interior with carbolic acid or silver nitrate will usually suffice.

ANEMIA OF THE SKIN.

Cutaneous anemia is most usually a part or symptom of systemic anemia. It is characterized by pallor and coldness of the general integument. Anemic conditions due to psychic excitement, as anger, or to reflex action from the digestive tract, as occurs in malaise, colic, etc., also local anemias due to cold or to transitory occlusion of larger vessels, are of no importance, as they are only of short duration and are not followed by further changes in the skin. Of more importance, as far as the final result is concerned, are the local and universal anemias of the skin, already referred to, when they are long continued or exert their influence frequently at short intervals, as they lead to interference with secretion and nutrition. The skin becomes dry and the epidermis exfoliates in lamellæ. The skin becomes lax, and atrophic conditions, excoriations in places, and even deeper necrotic ulcers may result.

HYPEREMIA OF THE SKIN.

Of greater importance are the cutaneous hyperemias. They are due either to congestion of blood in an irritated area of the skin (active hyperemia) or to stasis when the return-circulation is interfered with (passive hyperemia or hyperemia due to stasis).

Active hyperemias (*erythema congestivum*) are the result of engorgement of the smallest capillaries in the papillary layer. Large areas on the surface of the skin are pale red or bluish-red. Frequently the redness appears in small circumscribed spots and disappears on slight pressure, to return as soon as pressure is withdrawn.

3

Sometimes patients feel a slight itching or burning. Such hyperemias, of a rapid transitory character, disappear without causing any change in the skin. When long continued or of frequent recurrence, however, they lead to desquamation of the epidermis, accumulation of pigment, and to increased activity of the sebaceous and sudoriparous glands.

These hyperemic conditions arise from mechanical, thermic, or chemic irritants which come into direct contact with the cutaneous surface. Peripheral irritation—*e. g.*, scratching—may also be conveyed by reflex to other central nerve-tracts and give rise to a hyperemic condition in remote places of the surface. Finally, psychic disturbances—*e. g.*, shame and other psychic emotions—may cause direct irritation of the vasomotors from the cortex of the brain and thus produce hyperemia.

Livedo belongs to the stasis-hyperemias. It is due to interference with the return-circulation by the pressure of a bandage or tumor on the returning veins, to cold, or dilatation following inflammation of veins; larger or smaller areas of the skin show a bluish discoloration.

Cyanosis is a more widely distributed bluish discoloration of the skin, usually associated with dilatation of the vessels. It is due to occlusion of the larger veins, or directly to cardiac lesions or to stasis in the larger vessels. These conditions bring about the permanent changes leading to various consecutive processes, as chronic edemas, thickening of the skin, etc.

DERMATITIS.

Inflammatory processes in the skin are preceded by hyperemia. When an irritant is applied to the vasomotor nerve-branches, alteration in the vascular capillaries occurs and active hyperemia results. This is the precursory stage of inflammation. It is but a short step from hyperemia to inflammation, at first almost imperceptible. When exudations and infiltrations and changes in the cellular elements have occurred—for example, proliferation of the cellular elements—inflammation becomes more decided. Although these fundamental principles of inflammation are always present at the same time, the clinical picture differs according to one or the other becoming more pronounced.

ERYTHEMA.

Under this designation we group those mildly inflammatory conditions of the skin occurring in the most superficial layers and accompanied by slight or moderate exudation.

Erythema multiforme (Plate 6) is the type of this class. Vascular dilatation, active cell-migration, and an edematous saturation of the papillary layers, and also moderate proliferation of the connective tissue form the substratum of the cutaneous inflammation. Proliferation of the epidermis-cells in the rete and loosening or bullous elevation of the epidermis complete the picture of this inflammatory process.

Erythema multiforme most usually appears on the forearms and upper part of the arms, over the ankle- and knee-joints—in fact, over the extensor surfaces of the extremities; further, on the face, neck, nucha, and chest. Papules crop out rapidly; these spread, and in a few hours become deep-red patches. In a few days the skin is seen to be covered with macules and papules, projecting above the level of the skin; the older lesions are depressed in the center and begin to fade, but extend at the periphery with a red margin. Adjacent efflorescences coalesce and form with the successive crops the type of polymorphism. When the condition does not advance beyond the first stage of development we have an *erythema papulatum*. When papules and macules of the same age are principally present and confluence predominates they form the so-called *erythema gyratum, erythema figuratum.*

According to the amount of exudation characterizing the process, there **occur** elevations **of** the epidermis in the form of vesicles, the size of a lentil to that of a pea, which are situated **on a red** base and are tense and firm —*erythema vesiculosum, erythema multiforme bullosum* (Plates 7 and 7, *a*).

When these erythematous **spots, or the** vesicles, are **arranged in** rings—*i. e.*, when a new ring appears around one or more **of** these macules or **bullæ**—it constitutes respectively the so-called *erythema iris* and *herpes iris*. When the older vesicles desiccate in the center and **the** new peripheral ring alone remains we **call** this form *herpes circinatus.* These two latter forms occur principally on the backs of the **hands and feet**; they are usually associated with erythematous patches and rarely in fact appear independently of these. The outbreak lasts for two to several weeks, and frequently recurs at the same time **the** following **year.**

The course of the efflorescences differs according to **the amount of exudation**; they **may fade** in eight **to ten** days, disappearing **with** accompanying slight desquamation **of the** epidermis. They may, however, remain four to **six** weeks; and especially **when** they appear in successive **crops,** which is usually characteristic of **the** disease, they **may** annoy the **patients** for several months. At times **the** mucous **membranes** of the oral **cavity** and genital **tract** participate in the disease.

In addition **to** these objective phenomena the process is accompanied by moderate itching, **at times by a** burning sensation, languor, **and** psychic depression. In accordance with other observers **we** could often demonstrate troublesome gastric disturbances in our cases. At times patients complain of pains in the joints, which may develop into aggravated articular affections. Of **rare** occurrence are albuminuria **or** hemorrhages from **the** kidneys, and inflammatory complications of serous **membranes,** conditions which must be regarded as being due **to a high** degree **of** general intoxication. Usually a slight rise of temperature is noted, sometimes even high fever, which, however, does not follow any certain type.

Another type of erythema, which is distinguished from the ordinary erythema multiforme only by its external form and not by its character, is *erythema nodosum*. Frequently both forms appear side by side. In erythema nodosum intensely red and usually somewhat deep-seated nodules (Plate 8) appear over the extensor surfaces of the tibia, knee- and ankle-joints, more rarely over the articulations of the hands and on the forearms. The nodules increase in size, several may fuse together, and the affected parts frequently show a marked increase in volume. The nodes are very sensitive to pressure, and have a hard, elastic feel. The accompanying general disturbances are *mutatis mutandis* the same as in erythema multiforme: Nausea, feeling of weakness, fever, and articular pains. The swelling declines in one to two weeks and the entire process is usually over in about a month. Hemorrhagic infiltrations not infrequently occur in these nodose swellings along with the serous exudation, they turn bluish (*erythema contusiforme*, Plate 8), undergo gradual involution and show the well-known changes of color from green to greenish-yellow.

As to the causes of erythema and related processes, we are up to the present date forced to look to a few empiric facts and more or less theoretic suppositions. Experience teaches that certain kinds of fruits—*i. e.*, strawberries, raspberries; further, oysters, crabs, lobsters, seafish; especially fat, stale pork or sausage—may give rise to digestive disturbances and to erythemata. According to our clinical observations, made years ago, it is not difficult to imagine that after a certain cause—*i. e.*, eating of damaged food—not only the substances referred to above, but also others that are formed in the organism from imperfect digestion, produce various disturbances, especially in the digestive tract, to which is added the erythema and the clinical picture completed. Our co-laborer in chemistry, Dr. Freund, could always in such cases demonstrate a considerable number of toxins and ptomains in the excretions. Unfortunately, experimental proof is lacking to explain these processes thoroughly.

Treatment.—As erythema multiforme can be demonstrated to be, or at least may be supposed to be, of probable intestinal origin, the diet should be correspondingly regulated ; and, when indicated, laxatives and intestinal antiseptics should be resorted to. As such, we prescribe either menthol (0.2 (gr. iij) per dose in gelatin capsules, t. d.), or—

R̥ Pulv. cort. cinnamomi,   0.20 (gr. iij);
   Ol. menth. pip.,
   Ol. eucalypti,   āā gtt. j–ij.
Ft. capsula. (una).
Sig.—Four to six capsules daily (Freund).

General treatment is, in other respects, according to the usual rules ; rheumatic pains and articular swellings, etc., when present, are treated with local applications (ice-water, plumb. acet. bas. solut., Burow's solution), and internally salol and sodium salicylate are given.

When there is tendency to itching it is well to have the affected areas painted with spirituous solutions of carbolic acid and spirituous solutions of salicylic acid, etc., about the same strength as advised in hyperidrosis (*q. v.*), followed by dusting with starch.

The same therapeutic remedies will suffice for erythema nodosum and purpura rheumatica (Plate 11).

Several similar processes are allied with the typical erythemata, which they resemble partly etiologically, partly clinically—*i. e.*, as far as their external course is concerned. Urticaria comes first ; it is characterized by the rapid appearance of wheals, elevations which are frequently pale red, rarely white, and are surrounded by a hyperemic halo. As they disappear and reappear rapidly, and here and there become confluent, it is scarcely possible to state their size, because stable efflorescences are not found frequently. The spots seldom project more than 1–2 mm. above the level of the skin. Owing to paling of the center and peripheral extension of the process, urticaria presents the picture at times of a serpiginous affection. In some instances urticaria is closely allied to erythema multiforme. In urticaria as well as in the erythemata the appearance of edematous swellings, the cropping out of bullæ, and participation of the mucous membranes are of not rare occurrence.

Urticaria is especially characterized by severe itching, which is exceedingly troublesome to patients, as it robs them of sleep; and if the disease is of any duration, they grow weak in consequence of imperfect rest and the nervous tension. The itching leads to scratching, and this gives rise not only to localized new eruptions, but is also conveyed by reflex to remote and more extensive cutaneous surfaces. The skin of certain persons who are predisposed to these erythematous eruptions is so very sensitive that every local irritation is followed by an eruption of wheals (*urticaria factitia, autographism, l'homme autographe of the French*). Such individuals are frequently nervous and hysterical.

Worthy of special note are those forms of urticaria which appear in childhood, and which, owing to their persistence and frequent recurrence, trouble the patients for many years and leave brownish pigmentations behind (*urticaria pigmentosa*).

If any one of the erythematous diseases is entitled to the term *angioneurosis*, it is urticaria, because it presupposes a nervous disposition, inasmuch as slight peripheral irritations are frequently followed in a very short time by urticarial eruptions on remote parts of the body. The external irritants may be the bites of fleas, lice, bedbugs, gnats, or stinging-nettles. Urticaria is also met with in prurigo, in pemphigus, in pruritus of diabetics and jaundice, and likewise in disorders of menstruation and puerperal diseases with more or less pronounced participation of the uterus, as flexions, pregnancy, etc.

Further, the ingestion of foods and fruit is to be mentioned, as has already been done under the head of erythemata.

Treatment.—When an urticarial eruption cannot be attributed to external irritants (epizoa), the condition of the general health, and especially of the intestinal tract and the genital system, should be given careful consideration. In some persistent and recurring cases it will be necessary to regulate carefully the diet, or, when indicated, to combine a bath-cure (Karlsbad), and in other cases to treat any existing disorder of the digestive or generative organs.

For internal treatment are recommended: Arsenic, atropin, ichthyol (0.2 (gr. iij) per dose, Lang), antipyrin, salophen (4–5 (gr. lx–gr. lxxx) per day, de Wannemaeker), calcium chlorid (0.2–0.3 (gr. iij–gr. ivss) per dose t. d., Wright).

Brocq advises:

R, Quinin. muriat., 0.05 (gr. $\frac{3}{4}$) ;
   Ergotini, 0.05 (gr. $\frac{3}{4}$) ;
   Extr. belladonn., 0.02 (gr. $\frac{3}{10}$) ;
   Glycerini, q. s. ad pil. unam.

Sig.—Eight to sixteen pills daily.

Locally: Applications of the spirituous lotions already mentioned. The following may also be used:

R, Spirit. lavandulæ, 100 (f℥xxv) ;
   Spirit. vini gallici, 150 (f℥xxxviiss) ;
   Æther. sulph., 2.5 (gr. xxxiij) ;
   Aconitini, 1 (gr. xv).—M.

Sig.—To be painted on.

R, Acid. salicylici, 1 (gr. xv) ;
   Acid. carbolici, 2 (gr. xxx) ;
   Glycerini, 50 (f℥xiiss) ;
   Spirit. vini, 100 (f℥xxv).—M.

Sig.—To be painted on.

Further, baths containing starch, alum, **corrosive sub-**limate, washing with vinegar, etc.

Allied to urticaria is *œdema cutis circumscriptum, angioneurotic edema,* described **by** Quincke **and** others. In this **somewhat** rare disease edematous, **cutaneous** phlegmonous swellings appear, which may **be** the size of the palm **of the** hand, and **which** gradually merge **into** the normal **skin.** They disappear in one **place, to** reappear **soon** upon **some** other portion **of** the **body.** The **mucous** membranes of the mouth, pharynx, **and** larynx are **also** frequently implicated. Vomiting **and** local disturbances, **due to** swelling of the mucous membrane, are the most annoying concomitants **of this** affection. Riehl regards this morbid condition **as an** angioneurotic disturbance of the circulation similar **to an** urticaria.

Another diffuse erythema appearing symmetrically **on** the hands or feet **is** *erythromelalgia.* Patients **at first** complain of attacks **of** burning and pain, which are succeeded by erythema of varying intensity, which, however, **persists** for some time. This process is also regarded as **an** angioparalysis. Other observers attribute **it** to **central** pathologic **processes** in the nervous system.

Erythemata occurring in infectious diseases (**toxic** erythemata in the **stricter** sense). In connection with the above-mentioned erythematous and slightly inflammatory cutaneous diseases we would call attention to **those** pathologic products on the skin which precede or accompany **various infectious** diseases. In enteric fever, cholera, grave pneumonias, septicemia, acute exanthemata, etc., not infrequently small spots of roseola are observed on **the** trunk, **most** usually on **the** epigastrium and **on** the flexor surfaces **of** the extremities, and also extravasations of blood in **the form of** ecchymoses and petechiæ. **These** phenomena occasionally may be due directly to microorganisms collecting in the capillaries; the explanation, however, attributing them to intoxication affecting **the** nerves of the vessels, is much more plausible.

PELLAGRA.

Pellagra, mal rosso, mal del sole, in its early stage appears as an erythematous malady, which during its further progress exhibits the anomalies of pigmentation. Its supposed cause (an intoxication) associates it closely with the erythemata.

In some regions (as **Lombardy,** Venetia, **Eastern Friaul,** Bukowina, Roumania, etc.) pellagra occurs endemically. It appears in the spring and summer at **first** as an erythematous skin-affection, which becomes dark brown; the eruption shows itself on those uncovered portions most exposed to the rays of the sun, as the face, the dorsal surfaces of the hands, and, in the peasants who go barefooted, on the dorsal surfaces of the feet also. Patients feel weak and suffer from a feeling of pressure in the epigastrium and frequent diarrhea. Desquamation of the epidermis occurs. The discoloration of the skin disappears in the winter, to reappear the next summer. Later the pigment turns darker and bluish-red and the skin becomes sensitive. Patients complain of chilly sensations and cold. Muscular weakness, anemia, despondency, stupor, and melancholia develop. A fatal issue is brought about by aggravated diarrhea, diseases of internal organs, and delirium.

The disease is attributed to an excessive diet of maize; damaged cornmeal especially is said to give rise to pellagra. Neusser is of the opinion that the poisonous principle is developed in diseased maize under the influence of the *Bacteridium maïdis,* and that it produces the disease in field-laborers who are debilitated by insolation and gastric derangements. According to this writer, pellagra is a chronic systemic disease, characterized by disturbances of delicate nerves in the domain of the sympathetic and its central nerves and arterial channels, caused by a toxic principle forming in the intestines of individuals affected and leading to autointoxication.

Treatment.—This is mainly one of diet. Nourishing

food, out-door life, and administration of iron-preparations are indicated. Advanced cases are not influenced by such measures and a fatal result is inevitable.

DRUG-ERUPTIONS.

In the majority of instances diseases of the skin due to the ingestion of drugs belong to the type of erythemata, but differ from these principally in being polymorphous. In common with most erythemata, they are accompanied by gastric disturbance and not infrequently by fever. It is of practical importance, however, to study these various skin-manifestations separately.

All drugs do not give rise to cutaneous eruptions; and individuals differ materially in susceptibility, many indeed being free from such influence. Lewin's statements (*Handbook of Pharmacology*) are interesting: Among 402 drugs he found that 204—*i. e.*, 50.7 per cent.—might possess the property of irritating the skin. Such action from drugs requires a *temporary or inherent individual predisposition*. Some patients have an idiosyncrasy for certain drugs and react to the smallest doses; others can bear larger quantities and also a larger application of a drug without experiencing unpleasant consequences of any kind.

The rapid appearance of a generalized eruption from drugs in certain individuals is often surprising, which can be explained only on the basis of reflex action; for scarcely has the drug reached the digestive tract before the exanthema is noticeable on the skin.

It is somewhat different in those cases due to local application, when the skin is irritated directly by a remedy which is taken up by the skin, not only giving rise to irritation of the area or areas to which it has been applied, but also by reflex action leads to similar eruptions on other parts (Plates 14, 23, 23, *a*, 24, 25, 25, *a*). Exanthemata are due more frequently, however, than was formerly thought to be the case, to absorption of materials by the blood, which are then excreted by the glands of the skin, and during their passage give rise to the cutaneous eruption.

It would be beyond the scope of this work to consider individually the numberless drugs which may produce irritation of the skin.

Erythematous and at times vesicular efflorescences of various degree are observed to follow the use of antipyrin, atropin, chloral hydrate, balsam of copaiba, opium and its derivatives, strychnin, sulphonal, turpentine, etc.

Arsenic is of special interest, for the reason that it undoubtedly has a closer relation in its action to the skin, and is, moreover, frequently employed in dermatologic therapeutics. It produces erythemata, edema (especially of the eyelids), papules, bullous eruptions, zoster, and pigmentary deposits.

Mercury in all forms and methods of employment may irritate the skin. Not to mention the countless cases which show a diffuse erythema after the application of a mercurial ointment, we meet with erythemata following its internal administration and after hypodermic mercurial injections; even after transitory external use of corrosive sublimate—*i. e.*, washing out a furuncle—we have noticed the occurrence of erythemata and even of eczemas.

The preparations of iodin, especially potassium iodid, and potassium bromid, cause acneiform cutaneous efflorescences to appear, which will be further discussed in the section on acne.

HEMORRHAGIC ERUPTIONS.

Partial hemorrhages into single nodules have already been alluded to when discussing erythema contusiforme and septic erythemata. In the following pages those diseases which are principally or extensively accompanied by hemorrhages will be considered (Plates 9, 10, and 11).

Peliosis, or *purpura rheumatica*, belongs here. It may occur simultaneously with varieties of erythema in the same individual. It differs from the ordinary varieties of erythemata in involving the joints more markedly, and the efflorescences over the articulations are more numerous.

Dark-red to blue spots, the size of a lentil to that of a pea, develop at first over the joints, later on the rest of the body, more especially, however, on the limbs; the lesions are situated on a level with the skin and rarely project above it; they do not disappear on pressure and soon assume a purple hue, or in very grave cases, owing to marked extravasation of blood, they are of a bluish-black color. Patients are prostrated and complain of pains in the joints. In many cases the joints are decidedly swollen, the exudation is serous, sometimes hemorrhagic. Moderate rise of temperature in the evening, languor, anorexia, and a feeling of thirst are constant concomitants.

The exciting cause of peliosis, despite the numerous investigations of late years, still remains unexplained. The hemorrhage may be preceded by hyperemia and stasis, usually of long duration. The blood escapes through the walls of the vessels by diapedesis; it is rarely possible to demonstrate capillary disease. Some observers have stated that hyaline degeneration, fatty changes in the endothelium, and formation of thrombi during this process take place. This state of affairs, however, would probably be found to exist only in petechiæ occurring in the course of grave diseases (tuberculosis, Bright's disease).

It is highly probable, however, that toxins and ptomains circulating in the blood either change the latter or cause angioparalysis of the smallest branches by influencing the vasomotors. As far as the changes in the blood are concerned, it is certain that the percentage of hemoglobin is greatly diminished. Microcytes and poikilocytes are found occasionally in fresh blood; and, further, the eosinophilous cells are increased in number. Here and there it has been possible to demonstrate microörganisms.

The older efflorescences undergo the ordinary changes of blood-coloring matter and appear greenish-yellow to reddish-brown. When hemorrhage into bullæ (in erythema bullosum) has taken place, they dry up into brown scabs. The process generally lasts four to six weeks, and tends in some instances to recur.

Morbus maculosus Werlhofii, or *purpura hæmorrhagica*, is a disease which is differentiated from peliosis by the number and extensive character of the hemorrhages.

In this affection irregularly generalized, scattered petechiæ and vibices appear over the entire body. The mucous membranes of the mouth and pharynx participate more frequently in this process than is observed in purpura rheumatica. Edematous swellings accompanied by hemorrhages occur, and when they involve the larynx they may cause dangerous symptoms of suffocation. Still graver complications are the occurrence of hematuria and endocarditis and pericarditis, conditions which go to confirm more fully still the intoxication of the entire organism.

In order to complete the subject of hemorrhages in the skin we will briefly refer to *scorbutus* (scurvy), which differs from morbus maculosus only in degree and is characterized, along with the phenomena peculiar to that disease, by involving the gums and the mucous membranes of the oral cavity at an early date. The gums are of a dirty-gray color, very loose, and undermined in places by hemorrhages.

Owing to necrosis of the mucous membrane of the mouth there is very pronounced fœtor ex ore. The hemorrhages on the trunk and extremities, the soft parts being permeated by larger extravasations of blood and forced apart, are of graver importance.

Scurvy and morbus maculosus Werlhofii, as experience teaches, especially the former, result from malnutrition in general and lack of fresh meat and vegetables, and occur most frequently in convicts and seafaring-men.

We would mention finally that *bleeders' disease (hæmophilia)* is a permanent inherited **tendency to** hemorrhages, and **is** often found to exist in fat, well-nourished individuals, whereas the affections discussed above are acquired diseases accompanied by disturbances of nutrition.

ACNE.

Eruptions **which** are situated principally **on** the **face,** and which upon superficial inspection present a similar appearance, have heretofore been included under the general **term** of **acne.** Formerly *acne vulgaris, acne rosacea,* and *acne mentagra* (sycosis) **were** discussed together, although each disease depends on **a** different pathologic **process.**

At the present day **we** designate **as** acne a disease consisting essentially of **an** inflammation **of** the **sebaceous** follicles.

It may depend **upon various** causes. In many instances the irritation **of** the cutaneous follicle and **resulting inflammation are** due **to** external noxious influences. Not infrequently we **must** seek the predisposing cause in the organism itself—*e. g.,* cachexia, debility. Finally, **we are** acquainted with substances which during **their** excretion from the body through the **skin** give rise **to** folliculitis. Some authors would regard staphylococci as the **cause** of some varieties (blepharitis ciliaris, hordeolum). Acne correspondingly presents different clinical pictures and does **not** always pursue **the** same **course.**

Acne vulgaris, **or** acne, **appears on** the **face** (nose, forehead, chin, and cheeks), **on** the **chest, and** on the back **(Plate** 30). Both **sexes are** attacked **alike.** Chlorotic, **anemic** girls are especially predisposed; **also** boys, probably **more** than girls, during the period of **puberty** (sixteen **to** twenty years), when the beard begins **to** grow. Digestive disturbances, such as habitual constipation, indiscretions in diet, etc., are frequently mentioned as causes. **We cannot** up to the present time offer **a plausible** explanation **for this** frequent complaint. We would, however, not like to be considered as regarding the above-named disturbances as entirely without influence in producing this disease. In such individuals the secretory activity of the sebaceous glands is noticed to be increased; very frequently seborrhœa oleosa is also present. The real cause of acne, however, is interference with free excretion by sebaceous plugs or comedones forming in the outlets of the sebaceous glands and follicles; this leads to swelling and inflammation of the follicles and the *neighboring* surrounding tissue; the black plugs can be usually seen in the middle of the papules (*acne punctata*). Where the sebaceous glands are more numerous, as on the forehead, the nasolabial folds, and chin, acne-papules frequently make their first appearance, and are usually more numerous here throughout the course of the disease. Aggravated cases, with increased swelling and inflammation, take on a reddish-blue color and have a pustule in their center (*acne pustulosa*).

When the tubercles are hard, tough, and arranged in rows or closely bunched, as on the eyelids, it is called *acne hordeolaris*.

Acne varioliformis, acne necrotisans, is a special variety, appearing at the margin of the hair and on the hairy scalp. In this form the small papules and rapidly-forming pustules dry into a crust; after this falls a slightly-depressed cicatrix remains. This is regarded as characteristic of the affection. A further variety, in which the subjective symptoms consist mainly of burning and itching, has been designated *acne urticata* by Kaposi.

Finally, there remains to be mentioned that form of acne with accumulation of granulation-tissue; this appears principally on the nose, and is known as *folliculitis exulcerans serpiginosa*.

In this chronic disease, which frequently lasts for years, inflammation recurs with more or less intensity, and the swelling and pigmented markings may frequently lead to considerable disfigurement. In addition to the whitish, flat, and sometimes depressed cicatrices we also see raised

macules and elongate pustules which are still red **and in various** stages of evolution and involution; **and** alongside **of these** we **also** encounter inflamed tubercles of different **sizes,** making it difficult for an inexperienced observer to recognize the process as originating **in** the follicles.

The inflammation spreading **to** the sebaceous glands and extending to deeper structures, larger cutaneous **abscesses** frequently occur, which contain fluid and some-**times** inspissated pus.

Owing to its long duration the disease becomes a great trial to patients and repulsive and unpleasant to friends. The general health is scarcely affected.

In so-called *acne cachecticorum* (Plate 3) the case **is** different. It occurs in debilitated, marasmic individuals; it is usually more extensive, and frequently is found also on the body, and especially on the lower extremities. Follicular lesions of a livid color make their appearance, which exhibit a tendency to necrosis and to be converted into small superficial torpid ulcers. Occasionally lichen scrofulosorum coexists. Hemorrhagic effusion around the follicles and into the inflamed tubercles not infrequently makes the picture of cachexia more complete.

We are, finally, familiar with certain drugs, already referred to, which may irritate the cutaneous follicles and lead to follicular inflammation on those parts with which they come in contact. Such a substance is tar, which when used on hairy regions plugs the orifices of the follicles and causes *acne artificialis* (also called *tar-acne*). A similar con-**dition** is seen on the dorsal surfaces of the hands and fore-arms of factory-employees who handle dirty paraffin. *Benzin, creosote*, etc. are also looked upon as favoring causes. The ingestion **of** iodin and bromid preparations is also known to produce **acne.** Potassium iodid **and** sodium iodid not only cause **the** well-known catarrhal symptoms, occurring on the mucous membranes (coryza due to iodin), but also produce irritation while being excreted through the sebaceous glands, **in** consequence of changes in the seba-ceous secretion, giving rise to disseminated acne-tubercles,

4

not only on the face, but also frequently on the entire body. These tubercles and pustules are often accompanied by slight burning and pain. Extensive swellings of the follicles are rare; usually they are not larger than a pea; they involute without forming cicatrices, if they receive proper care and attention.

In *bromid-acne* the follicles are more markedly infiltrated, and it is less disseminated than iodin-acne; it is usually confined, moreover, to smaller areas of the skin; owing to the infiltration and inflammation becoming more extensive, the follicles may be converted into raised, irregular plaques, up to the size of the palm of the hand. The surface of these plaques seldom disintegrates; only small moist spots situated on a more or less intensely reddened and irregular raised base are formed.

The diagnosis of this last-named type of bromid-acne is often very difficult, as it presents few characteristics and may readily be confounded with vegetating syphilitic ulcers, or even with epithelioma. We have observed an instructive case of this kind on the lower extremity. An uneven, slightly raised, ulcerating surface covered with granulations presented itself for consideration. The patient, an aged female, had been taking large doses of potassium bromid in secret. The supposition that we had a bromid-acne before us, and not syphilis or epithelioma, was strengthened by the absence of symptoms pointing to syphilis, the presence of decided inflammatory phenomena, and also by the more rapid course than occurs in epithelioma.

Treatment.—Internal causes, chlorosis, disturbances of the stomach and intestines, and difficulties of menstruation are to be considered. These must receive their share of attention; and their management must go hand in hand with local treatment. The little pustules and abscesses are opened first; tubercles which may exist are punctured. When the small incisions and punctures have been healed by compresses or indifferent ointments and bandages the affected parts are thoroughly washed with soap and warm

water. Potash-soap, tincture of sapo-viridis, and the
legion of medicated soaps can be used. This treatment
suffices for many mild cases. Usually in connection
with the soap-washing, which is to be repeated at least
nightly, an ointment must be ordered. We mention :

℞ Sulphur. præcip.,
  Potass. carbonat.,
  Glycerini,
  Aq. laurocerasi,
  Spirit. vini gallici, *āā* 10 (ʒiiss).—M.
Ft. pasta.

℞ Sulph. lot.,   10 (ʒiiss) ;
  Balsam. peruv.,
  Camphoræ,   *āā* 2 (gr. xxx) ;
  Saponis viridis,   5 (gr. lxxv) ;
  Adipis,   30 (ʒviiss).—M.
Ft. ung. (Eichhoff).

℞ Bismuth. subnitrat.,
  Hydrarg. præcip. alb.,
  Ichthyoli,   *āā* 2 (gr. xxx) ;
  Vaselini,   20 (ʒv).—M
Ft. unguentum.

Sig.—To be applied thickly before bedtime (Hebra-
   Ullmann).

℞ Camphoræ,
  Acid. salicylici, *āā* **0.3–0.50** (gr. ivss–viiss) ;
  Sulphur. præcip.,   10 (ʒiiss) ;
  Zinci oxidi,   2 (gr. xxx) ;
  Saponis viridis,   1 (gr. xv) ;
  Ol. physeteris,   12 (ʒiij).—M.
Ft. unguentum.

Sig.—To be used externally every evening (C.
   Boeck).

Schütz recommends :

> ℞ Sulphur. lot.,
> Calcii sulphurat.,
> Calcii phosphat.,   *āā* 25 (ʒvj gr. xv).—M.
> Ft. pulv. subt.
> Sig.—Sulphur powder.

This is mixed with a little water and allowed to remain on during the night.

Further, lotions of :

> ℞ Sulphur. præcip.,   15 (ʒiij gr. xlv);
> Camphoræ,   12 (ʒiij);
> Aq. destill.,   250 (f ℥viij).

Or

> ℞ Sulphur. præcip.,   10 (ʒiss);
> Spir. vini Gall.,   50 (fʒxiiss);
> Spirit. lavand.,   10 (fʒiiss);
> Glycerini   150 (f ℥iv ʒvj).

> ℞ Sulphur. præcip.,
> Spiritus vini gall.,
> Aq. rosæ,   *āā* 30 (ʒviiss);
> Mucilag. acaciæ.,   10–20 (ʒiiss–ʒv).
> Sig.—To be used every three hours.

And other spirituous solutions and mixtures of similar composition.

We note very good results with **Lassar's** method of producing exfoliation :

> ℞ β-naphtoli,   10 (ʒiiss);
> Sulphur. præcip.,   40 (ʒx);
> Vaselini,
> Sapon. viridis,   *āā* 25 (ʒvj gr. xv).—M.
> Ft. pasta.

This paste is applied as thick as the back of a knife and is allowed to remain for fifteen minutes to one hour,

when it is wiped off and an indifferent powder is dusted
on.   The patient applies a 10 to 20 per cent. resorcin
paste, which is allowed to remain over night.   In a
few days inflammation of the skin, treated in this manner, results, the epidermis exfoliates and the acne is usually much improved or cured [?—Ed.].   When improvement alone results, this procedure is to be repeated.

Unna uses the following paste to bring about exfoliation :

℞ Resorcini,   40 (ʒx) ;
  Zinci oxidi,   10 (ʒiiss) ;
  Terr. siliceæ,   2 (gr. xxx) ;
  Adipis benzoinat., 28 (ʒvij).—M.

Ft. pasta.

SYCOSIS.

Synonyms : *Acne mentagra, Folliculitis barbæ, Sycosis
vulgaris.*

Sycosis is exclusively a disease of the hairy parts of
the body.   The ordinary and most common seats of the
affection are the hairy portions of the face, as the upper
lip, the cheeks, the chin.   The eyebrows and eyelids, the
nostrils, axilla, the pubes, even the hairy scalp, may in rare
instances show a similar follicular inflammation.

We have to deal with an inflammation of the follicles
and perifollicular tissue.   The first or primary stage of
the eruption consists of papules, which change into pustules and are pierced in the center by a hair.   These hairs
when pustulation is advanced and of some duration, are
loose, and on removal the sheath of the hair-root is seen
to be yellowish, infiltrated with pus, and swollen.   On
pressure with the finger-nails pus can frequently be made
to flow from the follicle.   When the pustules are crowded
together, larger inflammatory infiltrations result, which
are covered with crusts and scabs (Plate 31).   After the
scabs drop off a cicatrix may remain, the follicle is obliterated ; as a rule, however, in many cases of sycosis no

permanent **trace** is left. **In** long-continued sycosis, **or peculiar** forms of the affection, and when the disease has **extended over a** larger surface **and is of** the actively suppurative type, there remain cicatricial areas partially **or** completely **devoid of** hair (**lupoid sycosis**, ulcrythema sycosiforme).

This affection may persist **for** years, and as it attacks exposed portions, as the face, it is exceedingly annoying **to** patients. The pustules are furthermore sensitive to **the** touch and **very** painful when the inflammation is extensive.

We are unacquainted with **the causes** of this nonparasitic variety **of sycosis.** We only know that eczemas occasionally give rise to folliculitis, and that chronic nasal catarrh is **sometimes followed** by **sycosis** of the upper lip. [**In recent years** investigations of this disease point to pyogenic **cocci as the** essential etiologic factor.—ED.]

In **connection** with sycosis, **it** appears **to us** the proper **place to** refer briefly **to a** disease, described by Kaposi **as** *dermatitis papillaris capillitii,* which other authors (Bazin, Rogets) have called *acne-keloid.* Tubercles and tuberculo-pustules form at the margin of the nucha and posterior scalp; these develop into papillomatous vegetations, bleed easily, and are covered with crusts, and sometimes here and there contain pockets of purulent fluid. **The** process advances upward from the occiput to the vertex. The hairs are gathered **in** tufts or are entirely absent. New formation of sclerotic connective tissue, atrophy, and baldness result. **At times** tufts of hair protrude from the sclerosed tissue. **In** most instances the **disease tends** to limit itself **to** the lower occipital region.

Treatment.—It will be possible to retain the beard during the **period** of treatment only in mild cases. In aggravated **cases** the beard is cropped as close as possible and the crusts are softened with an emollient ointment. The hairs in the mature pustules are then removed with depilation-forceps and the beard is shaved. When the patient **is** very sensitive or when decided inflammatory reaction exists, it is frequently impossible, at least in the first week or two, to shave, and a paste of barium sulphid (depilatory) is substituted. Barium sulphate, charcoal, and linseed oil, according to Lestikow's directions, are stirred into a paste and subjected to a very hot coal fire; barium sulphid is thus obtained as a dark-blue powder. The following is ordered: Barii sulphidi, 10 (ʒiiss); zinci oxidi, amyli, *āā* 5 (gr. lxxv). The powder is made into a paste with water and is applied pretty thickly to the affected parts with a wooden spatula; in five to ten minutes it is to be washed off. Existing abscesses are incised and gray plaster is used to bring about **resolution** of tubercle-formation. Sulphur-paste, sulphur-mixtures, sulphur-soaps, and Wilkinson's ointment are adapted to further treatment. Furthermore, **washing the** parts cautiously with spirituous solutions of **corrosive** sublimate ($\frac{1}{2}$–2 per cent.), **resorcin** (5–10 per cent.), and pyrogallol (2 per cent.) are to be recommended.

Eichhoff has the following **solution well rubbed into the skin :**

    ℞ Naphtalini,
       Acid. **salicyl.,**   *āā* 3 (gr. xlv) ;
       Chloroformi,
       Spirit. **vini,**
       Glycerini,   *āā* 10 (fʒiiss).—M.
       Sig.—**To be painted** on.

Sycosis **of the** nasal mucous membrane and of the hairy scalp is treated **on** the same principles.

Similar methods in the main are employed in the treatment of parasitic sycosis (tinea sycosis, tinea trichophytina barbæ, *q. v.*). Ehrmann treats this **variety** with electric cataphoresis. The electrodes are **open in front or** contain receptacles **of** hard rubber, into which ichthyol (10 per cent.) is poured. The electrode is then applied **to** the skin and a **current of 15 to** 20 milliampères **is used** for ten to fifteen minutes.

ACNE ROSACEA.

Acne rosacea is characterized by red or bluish discoloration and hypertrophy of the cutaneous structures of the nose, and occasionally extends to, or is seated upon, other parts of the face, as the forehead, cheeks, and chin. This affection appears usually in adults, more frequently in men, but also in women ; in the latter exceptionally during the period of puberty, most frequently, however, during the climacteric.

Patients at first complain of a sensation of warmth in the nose upon the slightest cause, as when entering a warm room, excitement due to psychic irritation or to drinking ; at which time especially the nose appears flushed, which, however, soon disappears. The nose is observed to be frequently moist or oily—seborrheic. Sooner or later the redness becomes permanent and disappears only on mechanical pressure for a short period, to reappear as soon as this is withdrawn.

This intense redness goes hand in hand with slight or more or less pronounced swelling and hypertrophy of the nose. Occasionally a few venous vessels become more prominent at an early date. These dilated, tortuous varicose vessels impart a bluish color to the affected parts. The hypertrophy referred to is due to proliferation of the connective tissue, which begins around the vessels and is irregularly distributed. Frequently single flat papules develop superficially ; these increase in size and number, become confluent, and often form excrescences the size of a cherry to that of a nut. These lobular tumors, which are pedunculated at times, and the swelling of the nose may exceptionally increase to the size of a small fist, and the distorted organ overhangs the mouth (*rhinophyma*).

The skin of the enlarged organ is furthermore covered with dilated sebaceous follicles and scattered acne-papules and -pustules. At times patients also complain of burning pain, which is probably due to suppuration and formation of the acne-pustules. These enlarged noses retain their soft, elastic consistence for a long time, and only rarely feel tough and thick to the touch.

In the early stages the slight swelling of the nose may cause acne rosacea to be mistaken for lupus erythematosus; careful inspection will, however, prevent such an error. Soon the vascular alteration becomes conspicuous. The shiny, intensely red surface, the absence of being sharply defined from the surrounding neighborhood, and lack of scar-formation point to acne rosacea. The absence of disintegration and ulceration distinguishes acne rosacea from lupus vulgaris; the same applies to syphilis. Enlargement of the nose of higher degree in this disease, unaccompanied by excrescences, reminds us of rhinoscleroma; it differs from the latter, however, principally in being of softer consistence. [In the large majority of cases of acne rosacea met with in this country the condition consists of either diffused redness or additionally of dilated vessels and more or less numerous acne-lesions. Connective-tissue hypertrophy, except to a slight degree, is not very common. —ED.]

Popular opinion attributes the disease to drink. In most cases the abuse of alcohol must be recognized as the causative factor, sour white wine, whiskies, and brandies being regarded as especially injurious. These drinks, however, must not be looked upon as the direct cause; the chronic catarrhal conditions of the stomach and intestines of alcoholics must be regarded as the direct essential factors. Hence catarrhal diseases of these organs occurring in non-alcoholics may likewise be of similar etiologic importance. Experience, furthermore, demonstrates that individuals who are much exposed to cold—*e. g.*, coachmen, hucksters, and sailors—are frequently affected with acne rosacea. People of this class, however, are not very careful in their diet nor in the use of alcohol, and frequently resort to the latter for its warmth-giving effect. Excessive tea-drinking is also of causative influence. Our

observation, that such patients not infrequently have a
pale skin and conjunctiva, appears worthy of mention.

There are other etiologic factors to be considered. We
have mentioned that girls develop acne rosacea during
puberty and women more frequently during the climacteric period. Such individuals suffering from disturbances of the genital system are nearly always anemic.
We therefore may regard it as probable that long-continued anemic conditions dispose to this disease, and that
the anemia is the result either of digestive disturbances,
due to malnutrition, or to disorders of the genitalia and
loss of blood. A very hopeful prognosis therefore cannot
be given, in many instances, as the underlying causes may
be either difficult of recognition, or, when they depend on
the method of living, cannot be removed. The affection
never reaches a stage dangerous to life.

General treatment should be directed to the frequently associated symptoms of uterine disorders, abuse
of alcohol, disturbances of the stomach and intestines, and
constipation, which must receive proper consideration.

Schütz recommends the following as an intestinal disin-
fectant:

R Thymoli,   ~   0.4 (gr. vj).
  Solve in spirit. vini rect., 25 (f℥vj gr. xv).
  Aq. destill.,   150 (f℥iv ℥vss).—M.

Sig.—One tablespoonful in a glass of water at 10
       and at 5 o'clock.

To overcome constipation:

R Extr. aloes,   0.50 (gr. viiss);
  Ferri sulph.,   3 (gr. xlv);
  Extr. belladonnæ,   0.20 (gr. iij);
  Sach. et rad. liq., q. s. ad pil. No. 50.

Sig.—One pill t. d. after meals.

Local treatment, of course, is governed by the stage
and conditions. In the first stage applications of hot

water for a short period, covering the diseased skin with adhesive plaster or plaster of salicylic acid soap, and mopping with sulphur-lotions, will be productive of good results.

The following will be found serviceable :

R̠ Sulphur. præcip.,
    Ammon. muriat.,   āā 1.2 (gr. xviij);
    Spirit. camphoræ,   2.4 (gr. xxxvj);
    Acet. vini,
    Liq. cupr. ammon. mur., āā 4 (f℥j);
    Aq. laurocerasi,
    Aq. rosæ,   āā 15 (f℥iij gr. xlv).—M.
    Sig.—Shake and apply with finger (Schütz).

Application of tincture of iodin, iodized glycerin, and gray plaster will bring about absorption of hard infiltrations.

When numerous tubercles and dilated vessels are present it is best to scarify the skin. The choice of the instrument, of which there are quite a number, may be left to the individual taste; personally we prefer the most simple instruments. Some authors (Hardaway, Lassar) employ the electrolytic needle in place of scarifications. The treatment of rhinophyma is purely surgical.

VESICULAR AND BULLOUS ERUPTIONS.

HERPES ZOSTER.

The main representative of this group is *herpes zoster* (Plates 12 and 13). Its appearance is frequently announced by sensations of pain in the domain of the nerves in which the eruption is about to occur; or patients often feel only a burning sensation in the affected area shortly before the lesions appear. Slight inflammation and swelling of the skin in the region of one or more nerves ensue, and papules crop out on the surface, which in one to three days become translucent vesicles, varying in size from a grain of buckwheat to that of **a pea.** This condition may retrograde **and abort.** The **disease,** however, often continues to spread; **the** bullæ frequently attain the **size** of a bean and cover the entire affected areas of **the** involved region, with the exception **of** the red borders.

The contents, at first serous and transparent, gradually **become** turbid, and finally dry up into brown seabs. The inflammation **declines;** the pain becomes less intense or **ceases, or is limited to atypical recurrent** neuralgias, **which** annoy patients once or several times daily. The disease usually **lasts three to six weeks.**

This typical **course** differs **very** materially **in** some **cases;** extravasations of blood, **accompanied** by violent neuralgic pains, may impart **a** blue or **dark-red** color to the bullæ (*zoster hæmorrhagicus*). Not only the bullæ, but also the **tissue-base** (upper layer **of the** corium) are permeated **by hemorrhages.** The severest variety, known as *zoster gangrænosus,* **is accompanied** by high fever and **pain, and the** accompanying dark greenish discoloration **indicates necrosis** of the skin (Plate 12, single groups).

Zoster of an uncomplicated **type, as** already remarked, **gets well** in several **weeks and new** epidermis **is formed** under the scabs. In zoster gangrænosus the gangrenous **eschar** is separated by suppuration and an ulcerated surface results, which cicatrizes slowly **and leaves** keloidal cicatrices behind. After the objective phenomena **have** disappeared, patients **frequently, more** particularly those of advanced years, **complain for a** long time **of** anesthesia in the affected areas; more frequently, however, **of** neuralgias, paralyses, and trophic disturbances, manifested by atrophy of **the** muscles **and** sometimes by falling of the hair.

This disease, originating solely under **the** influence of the nerves, is usually unilateral and follows the distribution of single nerve-branches. The intervertebral ganglia have been found to be diseased, which, **as** we know, receive an anterior **motor** and a posterior sensitive root from the spinal cord. Consequently the most frequent form of zoster is one which follows the peripheral distribution of a spinal nerve. Of the cephalic nerves it is usually the trigeminus, in which the ganglion Gasseri plays the same *rôle* as the intervertebral ganglia, already mentioned, do in the spinal nerves. Besides this common etiologic factor, central diseases of the brain and spinal cord, especially diseases of the vasomotor centers, may give rise to zoster; *bilateral zoster* is attributed to this cause. Finally, the nerve-branches may develop a perineuritis in their peripheral distribution or irritability, due to pressure, and in this manner an herpetic eruption may ensue without the central part participating. In this case the herpetic eruption follows the ramifications and anastomoses of the peripheral nerves, and does not always adhere to the main trunks; consequently there occur completely isolated foci of herpes zoster, which are not covered by the main nerve-trunks.

The clinical pictures of zoster correspond to the localization and to the severity with which the nerves have been affected by the toxic influence. In thoracic zoster we notice the first eruption of vesicles at the greatest curve of the ribs, in about the posterior axillary line. The anterior pectoral portions usually follow. Vesicles in groups, corresponding to a small cutaneous branch, five to eight in number, invariably appear, and are developed in a certain place contemporaneously and in the same manner. The succeeding crops behave likewise, and we can frequently demonstrate fresh vesicles at the periphery along with central groups which are drying up. It is noted that the herpetic vesicles frequently do not adhere strictly to the region which the ramifications of the nerves seem to assign to them, and appear on the median lines or ascending or descending in the domain of neighboring nerves. The anastomoses of the cutaneous nerve branches (known to exist) alone can be held accountable for this.

Hemorrhages into the ganglia and inflammatory changes

in them, or when long continued leading to death of the
nerve-elements, diseases of foci of the brain or of the
spinal cord, cicatricial formation with remnants of pigment and preceding hemorrhages, lead to diseases of the
nerves or nervous system giving rise to zoster. Direct
causes are frequently traumatic in character, as an injury,
a blow, pressure on a nerve or ganglion by neighboring
organs—*e. g.*, exudations, inflammations, diseases of bones
(periostitis, exostitis), or carcinomata. Sattler has observed
toxic forms of zoster, especially in the domain of the trigeminus, follow carbonic-oxid poisoning; and Blaschko
and others have observed it follow arsenical administration. Malaria may also lead to neuralgia and zoster.

Beside these recognized causes, the etiology of a number
of cases of zoster is entirely unknown. Its epidemic
appearance, frequently associated with other acute infectious symptoms or diseases, appears to point to an infectious cause, which, however, still remains to be proved.

Zoster usually attacks adolescents and young adults,
old individuals less often, children infrequently.

Herpes zoster faciei et capillitii corresponds to the region
controlled by the trigeminus. In the domain of the first
branch of the trigeminus zoster occurs most frequently on
the eye, upper eyelid (nervus supraorbitalis), angle of the
eye (n. supraorb. et trochlearis) (Plate 13). Zoster cervicalis corresponds to the domain controlled by the second,
third, and fourth cervical nerves. The occiput, nucha,
neck, and region of the shoulders also belong to the cervical plexus. The region of the upper extremities is sup-
plied by the brachial plexus and by the first and third intercostal nerves. The region of the chest is controlled by
the intercostal nerves. The nates, abdomen, and genitalia,
and part of the thighs belong to the domain of the lumbar and sacral plexuses. The last supplies the skin of the
perineum, of the genitalia, and of the posterior surfaces
of the thighs, and the nates downward over the extremities to where the crural nerve begins on the thigh.

Treatment.—The affected areas are to be dusted with

an indifferent powder, or mild salves can be used ; when
the pain is severe, extract. opii, extr. belladonnæ, or
orthoform may be added. To control the neuralgic pains
sodium salicylate (4–6 grams (ʒj–ʒiss) per day), antipyrin,
pyramidon (0.3 per dose (gr. ivss) t. d.), chloral hydrat,
quinin. hydrobromat. (Wolff) ; occasionally hypodermic
injections of morphin must be employed to relieve the
torturing neuralgia of some patients. Scharff injects
Schleich's solution in the intercostal space, close to the
point of exit of the nerve :

R; Cocaini hydrochlor., 0.2–0.4 (gr. iij–gr. vi) ;
    Potass. chlorat.,   0.40 (gr. vj) ;
    Morph. hydrochlorat.,   0.05 (gr. ¾) ;
    Aq. destill.,   200 (f ʒviss).—M.

Sig.—**Liquor** anæstheticus Schleich.

HERPES FACIALIS ET PROGENITALIS.

The frequent **herpetic** eruptions on the **face and** genitalia do not follow the type of zoster. They are preceded
by slight itching, and appear on the mucous membranes
and neighboring skin and form groups of vesicles, each
vesicle the size of a pin-head to that of a lentil, situated
on a slightly reddened and somewhat raised base. Herpetic eruptions around the entire mouth, involving the
carmine of the lips and extending to the mucous membrane, are only infrequently met with, and in such instances only when catarrh of the cavity of the mouth
exists. Herpes around the nostrils is frequently associated with herpes **labialis.** This form of herpes occurs
almost **exclusively** in young subjects with slight catarrhal
affections accompanied by fever, coryza, and **bronchitis ;**
also in grave diseases of the respiratory **tract,** pneumonia,
and intermittent fever.

Genital herpes behaves in a similar **manner. In men**
it occurs most frequently on the prepuce (*herpes præputialis*), more rarely on the glans. Although of short dura-

tion, this disease often occasions diagnostic difficulties, and is of great importance to the physician, inasmuch as energetic caustics and strong remedies may convert it into a chronic, **torpid** affection, resembling infectious ulcers. Very frequently slight swelling and tenderness of the inguinal glands accompany herpes progenitalis. **In** women genital **herpes** is met with on the labia minora and majora, which are more or less swollen ; **we** have repeatedly **seen** herpes spread over the entire **external** genitals, **the perineum,** and the inner surfaces **of** the thighs as a very grave and painful **disease.**

The exact causes of these forms of herpes are unknown ; they are probably of nervous origin. Fright, excitement, and slight febrile disturbances at times give **rise** to herpes labialis and facialis. In **some** individuals herpes præputialis may **be** due **to** persistent erection, **and** also **may** show itself within **two or three days after** sexual intercourse.

Treatment of this herpetic **disease** consists in application of mild dusting-powders or salves. The **parts** should be protected ; caustics should be avoided.

MILIARIA.

Miliaria rubra et alba—heat-rash, or prickly heat—an eruption of very minute vesicles, accompanied by profuse sweats, and appearing on the trunk and extremities, at first has **a** red, later, when the epidermis becomes macerated and opaque, more of **a** whitish color (therefore the terms rubra et alba). The **contents of the** vesicles have an alkaline reaction. We meet such eruptions in field-laborers during the summer or in tropical countries, especially **at** the seaside **after** bathing in **salt** water, and sometimes in healthy individuals after long-continued sweating ; also frequently **in children** during the hot weather. **The** general health **is** not interfered with. Fresh-water baths and **keeping** the skin dry cause the disease soon to disappear.

Miliaria crystallina, or *sudamen,* makes its appearance in the most diverse infectious diseases, on the neck, trunk, abdomen, and the flexor surfaces of the extremities, in the form of perfectly clear minute vesicles, the size of a pin-head, and at times the size of a small pea. The affected regions are neither hyperemic nor inflamed, and have the appearance of being covered with dew-drops. Miliaria crystallina occurs during the puerperal process, in endo-carditis, enteric fever, etc. The vesicles do not change materially, being finally absorbed, the thin cover simply scaling off. Its occurrence and even successive crops are of slight importance; the causative or associated febrile systemic disease is mainly to be considered.

Miliaria epidemica is a rare disease of greater impor-tance. It occurs epidemically, and is ushered in by rigors and fever; the patients sweat profusely and are very dull. The skin of the neck and rump is covered with tubercles, vesicles, or pustules.

The entire aspect of the disease conveys the impression of its being due to general systemic infection, and this view gains in importance owing to the individuals devel-oping constant fever, dulness, and stupor, and frequently perishing. During the epidemic of 1892 observed in Carinthia 24 per cent. of the cases proved fatal.

Conditions of temperature appear to influence the origin of this disease; the epidemics occur principally during the spring and summer, when the atmosphere is warm and moist. Nothing characteristic is found at post-mortem; it is striking, however, that such cadavers de-compose very rapidly.

IMPETIGO HERPETIFORMIS.

This rare skin-disease has been observed, with but few exceptions, only in pregnant women and during the puer-perium. The eruption begins on the inner surfaces of the thighs and inguinal region, on the umbilicus and breasts, spreads over the whole body, and even appears on the

5

mucous membranes. Innumerable whitish vesicles of pin-head size, situated on a reddened, slightly swollen base, develop, whose contents become opaque and dry into a thin whitish crust. The eruption, which at first is confined to areas the size of a pea to that of a penny, spreads rapidly and in a few days larger regions of skin are invaded. The eruption extends in the following manner: A reddened and swollen zone appears at the periphery of a desiccating area or border, upon which new lesions form. Upon removal of the above-mentioned thin, dirty-white crust, newly formed epidermis is either found underneath or the skin is moist after the manner of eczema rubrum.

The gravity of the disease is indicated by the condition of the general health. The patients have continued or remittent fever and rigors; they are prostrated and have lost interest in everything; the tongue is dry; there are vomiting at times, stupor, and even delirium.

The **prognosis** is very unfavorable. Of fifteen cases, thirteen ended fatally (Kaposi). A pregnant woman, who passed through the disease after delivery and developed grave symptoms, came under our observation; she recovered so far as to be able to leave her bed; the fever, however, returned, and she perished rapidly, exhibiting signs of collapse. Post-mortem findings were negative, as in other cases reported.

The **etiology** of this disease is unknown. Inferring from the course it pursues, it may be regarded as an infectious disease allied to some erythemata and varieties of herpes and pemphigus.

Treatment is wholly symptomatic. In all cases so far observed it could not be demonstrated that the disease is influenced by any therapeutic remedies.

PEMPHIGUS ACUTUS.

As belonging to the bullous eruptions, the rare disease acute pemphigus is to be mentioned. Following short prodromal disturbances of the general health, the temperature often rises to 40° C., and pea-sized, perfectly clear vesicles, which increase rapidly in size and are scattered irregularly over the body, make their appearance. These rupture, the epidermis becomes dry and desquamates, and a slightly pigmented spot remains.

Similar successive crops occur for two or three weeks; the general symptoms improve and the disease terminates. At times gangrene of the skin in defined spots has been said to occur as a complication. We have only once observed a case of this kind, terminating with simple desiccation of the bullæ.

This disease appears to be of an infectious nature; this belief is strengthened, by the fact that the entire organism participates, the temperature especially rising rapidly, when only relatively slight involvement of the skin exists.

PEMPHIGUS NEONATORUM.

Pemphigus neonatorum is a disease which appears in the first or second week of life; the main symptom is the formation of bullæ, inasmuch as important disturbances of the general health are absent. The contents of the bullæ become opaque in one to two days; they grow flaccid and rupture. New red epidermis, surrounded by the remnants of the elevated epidermis, appears at the base. The localization of the disease differentiates it from pemphigus syphiliticus. The latter occurs on the palms and soles along with other evidences of syphilis on the rest of the body; the base and surrounding tissue are more infiltrated, this condition being entirely absent in the affection acute pemphigus.

PEMPHIGUS ACUTUS CONTAGIOSUS.

A disease in children, characterized by a bullous eruption, which is apt to occur epidemically after vaccination, has often been described under the name of acute contagious pemphigus (*dermatitis exfoliativa of Rittershain*).

This disease appears more as a diffuse **inflammatory** affection of **the** epidermis over large **areas of the body.** The epidermis desquamates or is elevated by **serum and** dries into **crusts.** Or finally—in cases of higher degrees—the epidermis **is** raised **in the** form of flat bullæ, branny desquamation **occurs, or** the epidermis is rubbed off. Underneath, the general integument appears red.

Riehl **has** recently discovered in one case a fungus with **long mycelial** filaments, and **also** regards this parasite as **the** causative factor in other exfoliative dermatitides. [It is generally believed that **some cases** of "acute contagious pemphigus," those **in which there are** scattered blebs of a benign character, are **examples of an anomalous type of** impetigo contagiosa.—ED.]

PEMPHIGUS.

We apply the title pemphigus, in the narrower **sense of** the word, to bullous eruptions whose course is characterized by an eminently chronic character. We differentiate **two** main **types,** *pemphigus vulgaris* and *pemphigus foliaceus.*

PEMPHIGUS VULGARIS.

The **far** greater number of pemphigus-vulgaris cases **must** be designated as **a** febrile disease, as they are ushered **in by rigors, rise of** temperature, nausea, and other disturbances. **Usually** outbreaks of erythema precede the eruptions of bullæ, **and** wheals resembling erythema annulare, figuratum, and urticatum, appear. Tense blebs develop **on these** wheals **or** erythematous spots. They may, however, **occur on** apparently **normal** skin without **being** preceded by other formations. **The** bullæ, which at **first** are the **size of** a pea, attain **the size of a** nut ; or when numerous and **close** together they become confluent and develop various irregular forms.

It is not so much the size **as** the number of bullæ appearing **on the** skin **at** the time of the eruption which characterizes a case as being of **more** or less gravity. **The** contents, at first serous and limpid, become opaque in a few days, the bulla ruptures, and the covering and exudate dry into a scab which is usually of a hemorrhagic character. In rare cases blood is in the earliest stage mixed with the contents of the bullæ. The inflammation is more marked where the bullæ and, later, the scabs cover large areas. The skin becomes hot and painful. Sometimes the disease is complicated by lymphangitis and adenitis.

Subjective symptoms are partly dependent upon impairment of the general health, thirst, anorexia, and marasmus being not infrequently associated; partly upon the processes on the skin, as burning, pains, tension, and itching, which interfere with **sleep.** The scabs gradually fall off and **a** young bluish-red epidermis appears underneath, which later on becomes pigmented and may remain **so for** varying lengths of time. Cases pursuing a benign course may terminate completely in two to six months, although such individuals may expect recurrences sooner or later.

There are, however, very mild cases of pemphigus **in** which the disturbances referred to are only observed in **a** minimal degree, **and** whose course is accompanied by only slight formation **of** bullæ. On the other hand, malignant cases occur in which numerous lesions appear and in which the above-mentioned systemic disturbances are very marked. In these latter cases the mucous membranes are also usually involved; and in such we may meet with bullæ and erosions having a whitish cover on the mucous membranes of the oral cavity, of the lips, tongue, palate, larynx, and pharynx, which are not only painful, but when involving the larynx may give rise to symptoms of suffocation (Plate 34, *a*). Pemphigus also attacks the conjunctiva and cornea. On the skin the efflorescences frequently pursue **a** different **course** from the one described —*e. g.*, the corium remains exposed after the covering of the bullæ has been lifted off or appears to be covered with a croupous exudate (*pemphigus crouposus*).

Pemphigus pruriginosus, as the name indicates, is characterized by severe itching, which interferes with sleep, and loss of strength, nervousness, and restlessness result. Owing to the lesions being destroyed early by scratching, excoriations, pustular eczema, extensive pigmentation of the skin, and melanosis result; in short, all the sequelæ belonging to chronic diseases accompanied by pruritus. [Many of these cases are now considered by the majority of American writers as belonging to the disease dermatitis herpetiformis.—ED.]

Neumann has called attention to a particular variety, namely, *pemphigus vegetans* (Plates 33, 34, and 34, *a*). These verrucous, ulcerating surfaces, depending on proliferation of the rete and papillary outgrowth, are thus formed : After the bullæ have broken the moist, oozing surface begins to be elevated ; the margins are raised in the form of flat, imperfectly raised bullæ and connect with the neighboring blebs; in this manner plaques the size of the palm of the hand are formed. The fungoid vegetations occur on the face, on the alæ of the nose and lips, on the ends of the joints, the genitocrural folds, the female genitalia, cleft of the anus and axillæ ; they pour out a secretion having a rancid odor and show a tendency to spread serpiginously. They seldom break down rapidly, but generally remain stationary for a long time.

Formerly this variety was regarded as certainly fatal, but owing to the modern method of treatment cases of late have remained alive for a longer period, as shown by the case depicted on Plate 33, and several others mentioned in the literature on the subject.

The papillary vegetations become flat when they are kept dry and disinfecting treatment is employed, and become covered with skin and cicatrize.

PEMPHIGUS FOLIACEUS.

Pemphigus foliaceus differs from the pemphigus varieties just mentioned by its more severe type and graver course.

This condition develops either after a long duration of pemphigus vulgaris, or quite flaccid bullæ appear from the first, whose cover is macerated and rapidly lifted off, leaving the corium denuded and red.

Owing to very deficient regeneration of the epidermis, we meet with large areas of excoriated epidermic lamellæ, which are partly covered with remnants of epidermis and are dried into thin crusts. Between the lamellæ the denuded corium or an imperfect epidermis appears. The scales are loosely adherent to the surface and exfoliate very readily (therefore, "foliaceus"). Owing to the gradual spread of the disease, the entire body-surface becomes affected. Irregular lines of skin denuded of its epidermic cover extend between the scales and exude serous fluid, which causes the clothing and dressings to adhere to the body. The hair of the entire integument is loose and usually falls out ; the nails are thin and brittle.

Patients experience great pain with every motion ; owing to fever and excessive diarrhea they become markedly emaciated and sooner or later succumb.

One form of pemphigus, as already indicated, may develop from another type of pemphigus. Usually, however, when the condition has lasted for years, we observe one form on one part and another on a different part of the body ; for instance, pemphigus pruriginosus and pemphigus vegetans (case shown in Plate 33), and pemphigus foliaceus, etc.

We therefore are led to suppose that the several varieties of pemphigus are only one disease.

The etiology of this usually ominous disease (according to Kaposi's estimate, 10 per cent. do not recover permanently) has remained unexplained up to the present. Post-mortem investigations have not developed anything tangible ; the individuals either succumbed to an intercurrent affection or to marasmus. [Comparatively few of the cases described in this country under the name dermatitis herpetiformis, which Kaposi contends are pemphigus-cases, are of a fatal character.—ED.]

72   *DISEASES OF THE SKIN.*

Bacteriologic examinations of the contents of the bladder and of the products of metabolism (urine) have also failed to furnish a positive explanation. As many nervous affections are known to be accompanied by skin-diseases with formation of bullæ, **pemphigus has been** attributed to this cause.

Occasionally, **in** some instances, **we are** enabled **by a** study of the cases to advance hypotheses attributing **pemphigus as a symptom of** another affection of the organism, appearing on **the skin.** Otherwise the etiology of **most** cases is enveloped **in** darkness, and such will be the **fact** until we possess a more intimate knowledge of the **disturbances** of metabolism and of the associated chemic and toxic processes in the organism.

Treatment.—As the entire organism participates in this disease, the general health **must** receive proper **attention** first. The strength must be improved by tonics, proper **diet, and** alcohol. Of internal remedies, **arsenical** preparations **are** worthy of most confidence, although their action in pemphigus must be said to be unreliable. When itching **is severe** the administration of calcium chlorid (1 gram to $1\frac{1}{2}$ grams per day (gr. **xv** to gr. xxiiss)) may be tried. Externally, **inert** dusting-powders, bandages, and ointments of boric acid and zinc oxid, and Wilson's ointment, etc., **are used.** When there is tendency to severe itching, painting and rubbing with tar and tar-ointments **are** indicated. When **large areas** are denuded of epidermis and considerable **serum** has been lost, and treatment **with** ointments, owing **to the general** condition of the patient, is difficult to carry **out** (pemphigus foliaceus), the use of the continuous water-bath is recommended; patients usually feel quite comfortable in **it.**

INFLAMMATORY DERMATOSES.

DERMATITIS.

In the section on erythema it was stated that the essential element of inflammation of the skin is hyperemia ; and in the beginning is, in fact, the only one. We have also referred to the superficial inflammations of the skin which are called forth by irritating substances (toxins, medicaments), showing the close relationship of, and the very slight differences between, hyperemia and inflammation. In the following brief summary we shall refer to inflammations of higher degree. These are caused either by pathologic processes in the organism or are the results of direct thermic, chemic, or mechanical injuries, to which the skin is often subject. As to the inflammations due to traumatic injuries, we consider such as belonging properly to the province of surgery.

Experience has taught that diabetics are predisposed to various kinds of cutaneous inflammation. Such individuals may suffer from anidrosis, asteatosis, pruritus cutaneus, sometimes erythemas, eczema, furunculosis, anthrax, and even diffuse dermatitis. Such dermatitides on the extremities occur as a result of slight pressure or slight injuries. It may easily happen, therefore, in such instances that the subcutaneous tissue of the soles of the feet, toes, ball of the foot, and dorsum of foot, become the seat of inflammation, which may lead to gangrene and bone-necrosis.

The pathogenesis of these conditions is not entirely clear. Kaposi holds the view that the sugar deposited in the tissues ferments, and thus gives rise to the inflammation. We might also call attention to the lessened resisting power of the organism as a factor in such patients ; especially as it is known that diabetics are not equal to much fatigue or to continued mental effort—in fact, their power of resistance and recuperation is much compromised. Experienced surgeons are well aware of this fact, and, if possible, avoid operations of gravity in such people.

Inflammations and even tissue-necrosis are encountered in enfeebled individuals after acute diseases, such as variola, typhoid, etc. In spite of the greatest care in some cases one is not able to prevent the formation of bed-sores.

In this same class, too, belong marasmic subjects and old men, in whom the circulation is weak (*senile* or *marasmic gangrene*); finally, cases in which there is contraction or closure of the arteries, as in atheroma of the vessels; in endarteritis obliterans, as sometimes observed in the distal arteries after syphilis, which leads to inflammation of the peripheral parts of the extremities and to progressive gangrene.

Finally, *multiple cachectic gangrene* is to be mentioned, which Simon and Kaposi have observed in enfeebled children, and thought due to capillary thrombosis.

In this group of inflammations of the skin are to be included these cases which belong to the domain of neurotic cutaneous disease. Among these the most important is *Raynaud's disease*, or symmetric gangrene, which has been observed in brain and spinal affections, and, according to Hochenegg, signifies a pure vasomotor disturbance, without primary disease of the vessels. The capillary vessels of the skin are contracted by vasomotor influence; there arises a local anemia; the skin feels cool and is pale. If the contraction lessens, there follows a congestion of the veins, which is characterized by regional cyanosis and swelling. If the circulatory disturbance is not equalized, if the trophic impulse is increased, there follow, with accompanying severe neuralgic pains and vesicle-formation, inflammations, and indeed gangrene, of the toes and fingers.

Paresthesias and anesthesias are observed in cases of symmetric gangrene preceded by markedly severe nervous disturbances.

Syringomyelia is also often associated or followed by

trophic disturbances of the skin; it is, however, to be distinguished from Raynaud's disease by the asymmetric appearance of the ulcerations, as well as by the appearance of various other eruptions in consequence of the dystrophy, as eczema, rhagades, panaritis, bleb-formation, gangrene, etc.

Perforating ulcer of the foot has been considered by some authors likewise as a trophoneurosis. It appears most frequently on the flexor side of the large toe and over the ball of the foot. Mostly a callous condition of the epidermis or a corn-formation precedes it. This accumulation is thrown off by underlying inflammation, and leaves an ulcer, which extends deeply and may even lead to necrosis of the bone.

Finally, we may refer to *spontaneous gangrene* in *hysteria*, of which young females are mostly the subjects. Preceded and accompanied with burning sensations, a quarter-dollar-sized to dollar-sized spot or a wheal-like efflorescence develops, which soon takes on a dark-blue color and is covered with a dry crust. This is cast off, the wound heals from time to time, and new gangrenous plaques appear. After a shorter or longer duration (up to two years) the eruptive tendency disappears.

COMBUSTIO (BURNS).

Under the term "combustio" are designated those cutaneous inflammations due to the action of heat or caustic chemic substances upon the skin. The tissues react after such injury in different degrees of inflammation, provided vitality has not been completely compromised or destroyed. On account of the frequency of this accident the skin inflammations in this group are of first importance. The most common cases in which burns are observed are from heated bodies or hot liquids, as hot pitch, hot water, petroleum, explosive materials; and of the chemic materials, mostly lime, caustic acids, etc. The surface of the body is always the seat of the first symptoms, although almost immediately thereafter also disturbances of a constitutional character present themselves. According to the effects produced, it is customary to divide burns into three grades, as follows:

1. *Burns of the First Degree.*—In this grade (combustio erythematosa) a small or large surface of the skin reddens with slight swelling, as observed in erythema, but diffused and not in wheal or papular form. This slight inflammatory condition of the skin is followed in a few days by a brownish coloration; it then gradually returns, with slight exfoliation, to the normal state. Generally, with this degree of combustio the general equilibrium is not disturbed, and the burning sensation of the skin is readily controlled by therapeutic measures.

2. *Burns of the Second Degree (Combustio Bullosa).*—The surface involved is the seat of vesicles and blebs from pea- to fist-size, tolerably well filled with serous fluid. The epidermis is not equally lifted up, as it is, for example, in pemphigus-blebs; and the covering is mostly thicker, the base of the blisters being the rete Malpighii or even the papillary layer. The surrounding skin is dark red and shining. The patient experiences a feeling of burning or heat in the part, which often extends beyond the immediate boundary of the burn itself. The smaller blisters remain unbroken, their contents becoming milky; the epidermis dries to dark crusts, which drop off and disclose the newly-formed epidermis. The larger blisters are torn upon removal of the burnt clothing or from pressure or contact in lying in bed, so that when first observed by the physician they are seen as irregular folds of epiderm or bared red areas. These are covered with whitish spots or specks, and in a few days become quite red, and are followed by an exudation and a cell-formation which gradually lead to complete over-skinning.

The subjective symptoms in these cases consist of marked pain and burning, which are heightened by the pressure in bed and by the removal of dressings or clothing. If a large part of the surface is involved, the life of the patient is endangered.

3. *Burns of the Third Degree* (*Combustio Escharotica*). —In these cases, in addition to the symptoms of the other grades, there is observed, as especially characteristic, a condition of mortification of the tissues, resulting from the intense action of the heat. The soft parts present, at least as to extent and depth of the burns, in every case all possible degrees. Most frequently the part, both skin and tissue, appears as if it had been cooked with steam or hot water. Very seldom are to be seen on the burned areas bullous elevations; but for the most part the skin is observed hanging in shreds. In other cases the affected regions present a mortification, in which the skin is whitish, alabaster-like, hard and tough to the touch, and lifeless in appearance. Worse still are those cases in which the skin and soft parts are converted into a dry, leathery, and hard dark-brown slough. The sloughs are irregular in area, and on the periphery symptoms of burns of the milder degrees are observed. In those unfortunate cases in which the body is exposed to direct flame the condition is one of carbonization and distortion. The patients are in the highest degree of agitation, and in this severe grade of burn succumb often in four to six hours (nerve-shock (Kaposi)).

More frequently, after a period of excitement there follow an apathetic condition, yawns, sighs, and gradually singultus, and even vomiting of gall. The patients grow restless, bewildered, are attacked with cramps and opisthotonos, lose consciousness, become delirious, and fall into a stupor. In these cases the bladder is found to contain but a small quantity of urine. The breathing becomes superficial, the pulse weak, and a fatal result soon ensues. If the patient survives the first two or three days, there begins a sharply defined inflammation with suppuration. The slough contracts and in the course of one to three weeks is cast off by the suppurative action. On the less-involved areas granulation begins. This

period is for the patient also a dangerous one, inasmuch as
he may suddenly die from heart-failure. Many authors
consider death in such instances due to a breaking down
of the red blood-corpuscles; others, to the formation of
toxic substances in the organism.

Irrespective of these direct dangers from the actual
burn there are other risks to the life of the patient later,
due to intercurrent disease, as pneumonia, Bright's disease,
erysipelas, and pyemia.

The scars following burns are often keloidal, hypertrophic, and cause in later years more or less difficulty;
the blood-circulation may be compromised, as a result of
which the peripheral part becomes enlarged by edema and
elephantiasic. Very often the movements of the head are
hindered by scars on the neck. Contraction of scars on
the extremities impairs the usefulness of the limbs, and
the arms are not infrequently drawn into fixed angles or
drawn to the trunk.

Treatment.—In burns of the first degree: Dusting
the parts with an indifferent dusting-powder, or ice-water
applications frequently changed, or aluminum acetate.

In burns of the second degree: Opening the blisters
and applying mild salves spread upon bandages. The
bared rete or corium is dusted with iodoform in a thin
layer, and over this a bandage of boric-acid salve or
dressings of equal parts of oil and lime-water. Von
Bardeleben recommends for the burnt areas solutions of
carbolic acid (3%) or salicylic acid (3%), and then to be
enveloped with soft gauze bandages which have been
previously covered with equal parts of bismuth and starch.
Such a dressing may remain on eight to fourteen days.

According to the latest experience, treatment with picric-
acid solutions has been commended. The burnt parts are
bathed for five to ten minutes with—

℞ Acidi picrici,   5 (ℨj gr. **xv**);
 Alcoholis,   80 (ℨiiss);
 Aquæ destillatæ,   1000 (Oij).

Immediately following this the wounds are covered with wadding or lint if the skin is still intact; or, if this is injured, then with sterilized gauze. These dressings are renewed every three or four days. In extensive burns compresses wet with this solution are kept constantly applied. According to our experience, this treatment, owing to its painfulness, is not to be recommended.

In extensive cases the continuous bath, according to Hebra, is especially serviceable. Internally alcohol is to be given; and if there is great restlessness, with loss of sleep, morphin, chloral hydrate, and the bromids.

Lustgarten recommends atropin, and lately Tommasoli the subcutaneous injection of artificial serum (that made of salt and sodium bicarbonate).

The management of burns of the third degree is to be according to the same general plan.

CONGELATIO (FROST-BITE).

Frost-bite arises after more or less prolonged exposure to low temperature. The time necessary for such action differs materially with different persons. Anemic individuals or those weakened by wading through snow suffer more severely than robust, healthy people. The appearances upon the skin are, as in burns, divided into the three grades—erythematous, bullous, and escharotic.

Frost-bite is most common in such parts as the uncovered hands, the poorly clad feet, the nose, ears, and cheeks. The patient experiences slight burning; soon loses, however, this feeling, and is only subsequently made aware that he has been frost-bitten by the thawing out, which is accompanied by sticking pain and intense itching. In this manner arises dermatitis erythematosa, so-called frost-bites, or *perniones*, or *chilblain*, appearing as variously sized, slightly raised spots of livid color. The blood-vessels become paretic, to which are due the bluish color, the serous infiltration, and the slight swelling. If these inflammatory appearances are followed by greater infiltration and exudation, the epidermis will be lifted into vesicles or blebs, the contents of which may be more or less hemorrhagic. Sometimes these give place to torpid ulcerations, which from their exposed situation heal slowly and may be troublesome through such complications as lymphangitis and adenitis.

As already mentioned, anemic individuals are especially exposed to this affection, especially the hands and ears; and in even moderate cold, after having once suffered from frost-bite, with its consequent blood-vessel changes, may readily be attacked again.

In extreme cases of frost-bite (congelatio escharotica) there always arise hemorrhagic blebs or a bluish, marble-ized, cold-feeling and insensitive surface. One cannot at first sight gauge the extent and the consequences in such cases, inasmuch as experience teaches that the soft parts, which may present the appearance of having been frozen, may yet recover, since the blood-vessels may remain permeable. In its further course a reactive inflammation occurs around the mortified areas; or after exposure to intense cold the mortification may not only extend through the soft parts, but even involve the bone.

Necrosis—casting off of the ear-lobes, or phalanges or entire fingers—is not infrequent. In these long-continued cases there is always the possible danger of absorption of putrid material, with consequent phlebitis and septicemia and death.

Treatment.—As already stated, anemic individuals are the frequent subjects of these accidents, especially of the first grade, on ears, nose, hands, or feet; it is therefore evident that in such cases the internal administration of iron-preparations is to be advised. Locally, painting with tincture of iodin, collodion, or the use of—

R, Acidi tannici,   2 (gr. xxx);
  Glycerini seu,
      Spiritus camphoræ, q. s. ad 50 (fʒiss).—M.

Sig.—To be rubbed in.

℞ Camphoræ **tritæ**,   3 (gr. **xlv**) ;
Lanolini,
Vaselini,   *āā* 15 (ℨss) ;
Acidi hydrochlorici **pur.**,   2 (gr. **xxx**).—**M.**
Ft. unguentum (Carrié).

℞ Bals. peruviani,   5 (gr. lxxv) ;
Misturæ oleoso-balsamicæ,
Aquæ colonienisis,   *āā* 30 (ℨj).—**M.**
Sig.—For external use (Rust).

℞ Calcis chlorat.,   1 (gr. **xv**) ;
Unguent. **paraffini**,   9 (ℨij gr. **xv**).—**M.**
Ft. unguent.

Sig.—Rub in a pea- to bean-sized piece five minutes and bandage (Binz).

Besnier and Brocq recommend bathing **with a solution of walnut-leaves** and painting **on**

℞ Aquæ rosæ,
Acidi tannici,   *āā* 0.5–1 (gr. viss–xv) ;
Glycerini,   30 (ℨj).

Then dust with salicylated bismuth powder (1 : **6**).

In acute cases it is advisable first to place the **person in** a cool room, **and to rub** the parts with snow **and adminis-** ter the **usual analeptica.**

ERYSIPELAS.

Erysipelas **is a disease due to** infection, **and is** always accompanied by systemic disturbance. The eruptive phenomena **may be** found upon **any** part **of** the body. The affected **area** is swollen, tense, and smooth, and **fiery** red. The disease not infrequently continues to spread peripherally, and in some cases it cannot be deter- mined beforehand how **far** the process will extend. The affected parts are tender **to touch**, and, especially on the peripheral zone, painful.

6

The disease does not invariably spread regularly from all sides, but sometimes shoots out in lines; or a neighboring part may be spared and it appear some distance from the original infection. Not infrequently it spreads along the lymphatics along the entire extent of the limb. A peculiarity is noticed in some cases, in that the disease heals at the place of first appearance and then spreads to the adjoining surface, extending in this way peripherally for some time and possibly involving a considerable surface (*erysipelas migrans*). Sometimes the parts already healed again become affected. In severe cases vesicle- and bleb-formation is a noticeable feature (*erysipelas bullosum*). In extreme cases the parts may even become gangrenous.

The Streptococcus erysipelatis (Fehleisen) is admittedly the cause of this disease. Inoculations of pure cultures of this microörganism have succeeded in producing true erysipelas.

The most frequent site for the disease is unquestionably the face. It often begins at the nasal apertures, in consequence of some exfoliation or fissure; or from the corner of the eye, or from some other point where there has been an injury or break in the continuity of the epidermis through which the infection gains a foothold, and then spreads out over the face, ears, and the hairy scalp, sometimes extending down the neck and possibly to the trunk.

Even before the redness appears there may be more or less fever and a feeling of being unwell; the temperature rises to 40° C. with every exacerbation. In cases involving the entire head the patient is soporific or often violently delirious; in those who drink freely—alcoholics —the disease is almost always accompanied with delirium tremens.

Experience teaches that recurrences are not uncommon, due either to the fact that some of the cocci remain in the tissues or that the disease arises from a new infection. Such recurrences frequently leave behind thickening of the connective tissues and elephantiasic enlargement; as, for example, on the lower extremities when in association with foot- and leg-ulcers. Falling out of the hair is a frequent consequence of erysipelas of the head.

Erysipelas heals with a lamella-like exfoliation of the epidermis or with the gradual dropping off of the crusts— the latter resulting from the dried-up blebs and vesicles.

It is worthy of note that the exanthems, as syphilis, psoriasis, and lupus, often disappear during the course of the fever in this disease (erysipelas salutaire of the French).

The **prognosis** depends upon the constitution of the individual, upon the severity of the attack, and especially upon the duration of the disease.

With erysipelas the so-called pseudo-erysipelas (*phleg-mon*) may be confounded. This phlegmonous inflammation usually has its origin at the seat of an injury, which either by immediate infection or subsequently is inoculated by septic material. Accompanied by chilliness and fever it may spread over an entire extremity—a thick, hard, painful, tense, and red swelling. Very seldom is there any tendency to retrogression; but usually pus-formation in the subcutaneous tissues takes place. Sometimes the process results in extensive purulent melting away of the tissue. The purulent action involves the fascia and muscles, often down to the bone. On opening a pus-collection great masses of bad-smelling pus mixed with tissue-débris are poured out. The patient, on account of the general infection and the severity of the local process, becomes emaciated and weak; and if he does not die in the acute stage of pyemia, he is endangered by the long-continued cachexia.

Treatment.—This consists of regulation of the diet, antipyretics, and alcohol. In investigating the source of inoculation, as, for instance, in facial erysipelas, inspection of the mouth and nose should be made, when it will often be found that the starting-point has been an abrasion from rhinitis or from a tooth-abscess.

Local poultices of aluminum acetate or lead-water, painting of the bordering healthy skin with iodin tincture, collodion, or ichthyol-collodium (10 per cent.), etc. We employ preferably the following salve applied on bandages :

R̶ Iodoformi,	30 (ℨj) ;
Creolini,	15 (ℨss) ;
Lanolini,	
Vaselini,	āā 30 (ℨj).—M.
Ft. unguentum.	

Of the many other remedial applications recommended may be mentioned absolute alcohol applied on compresses of lint, and which are wetted every fifteen or twenty minutes, over which are placed a dry cloth and gutta-percha tissue-paper (von Langsdorf) ; painting with guaiacol and olive oil, equal parts (Maragliano) ; oil-of-turpentine treatment after Luecke, in which rectified oil of turpentine is rubbed four or five times daily into the affected parts with a brush or a piece of lint.

FURUNCULUS.

Furuncle, or boil, is frequently observed to have its origin in an acne-pustule, or at least in an inflamed follicle. In the beginning there is noticed a painful inflammatory nodule in the skin. The apex gradually shows pustulation, in which sometimes a hair is found sticking. This pustule dries to a crust ; after three or four days the purulent infiltrated plug may be pressed out ; or this may be facilitated after the part has been linearly incised, the opening being thus enlarged. The cavity left closes gradually by granulation. It is repeated experience that boils are rarely seen singly, but that most frequently several appear simultaneously, or, what more frequently happens, many furuncles appear one after the other (*furunculosis*).

Boil-formations are often seen in connection with acne,

scabies, pediculosis, eczema, etc., the excoriations produced in these diseases by the scratching presenting favorable means of inoculation. The fact that boils appear successively on the same individual and close together, and also appear on several or more people living together, speaks strongly for the conveyance of the disease from one point to another and from one person to another (Plate 32).

Staphylococci have been recognized as the active etiologic factor.

CARBUNCULUS.

Carbunculus, or anthrax, appears most frequently at the nape of the neck, in the face, on the back, and in the sacral region. It is distinguished from furuncle by its larger size and painfulness. It is nut to small-fist sized, hard, very painful connective-tissue inflammation, which, after some duration, breaks through the surface at several points. Sometimes the overlying skin necroses to a dry leathery slough. The high fever and the intense painfulness make this disease one of some severity, and when the inflammation extends peripherally to any great extent there is grave danger that the patient may die from pyemia.

PUSTULA MALIGNA.

This is a disease of the cutaneous structures due to infection by the Bacillus anthracis of splenic fever, which, as known, occurs in animals, as horses, sheep, horned cattle, etc.

Inoculation occurs either directly to men who may be employed among animals, such as coachman, hostlers, and shepherds; or among those who have to do with the products from animals, as it is known that even hair and skins may convey the bacilli, especially the spores, which have an extremely tenacious vitality. The infection may take place through insect-bites, or by direct inoculation; less frequently through inspiring dust containing spores, or through the digestive tract from eating diseased flesh.

The course of such an infection is always dangerous, although it sometimes happens that the pustules, accompanied by mild systemic disturbance, finally heal, the infiltrated tissue being cast off.

EQUINIA.

Equinia, or glanders, arises most frequently through direct inoculation from affected horses or mules. The cause of the disease is a specific microörganism. It is known that remaining in an infected stall or handling affected animals exposes to possible infection.

The course of the disease is rapid, and frequently, in consequence of the chills, fever, pain, edema, suppuration of the joints, phlegmon, and gangrene, the individual's life becomes endangered.

Some cases may persist for years. In such instances, even when many nodes are present in the subcutaneous tissue with subsequent breaking down of the lymphatic glands, recovery may finally take place.

SQUAMOUS DERMATOSES.

PSORIASIS.

This disease frequently appears in individuals about puberty and early manhood. Children are comparatively seldom attacked; in advanced years it is met with, but usually as a recurrence or as a persistent eruption or remnant from the disease in earlier life. In the beginning the eruption consists of red papules, which in a few days show scaliness of a somewhat adherent character. The lesions increase in size and numbers, so that in the course of a few weeks the older efflorescences are to be seen as flat rounded scaly patches with a red halo, while recent lesions are seen near by.

According to the predominant form and size of the lesions, it is customary to designate the eruption by

various names.   In the early stages, when the lesions are
mostly of small size, as already mentioned, it is designated
psoriasis punctata (Plates **15 and** 16) ; if the plaques are
flat with considerable scaliness, it presents the so-called
psoriasis guttata, from its resemblance to the condition
which would be produced **by** sprinkling **a** handful **of**
mortar over the skin.   If the eruption consists of moder-
ate-sized, flat and scaly patches, it constitutes the most
common clinical variety—*psoriasis nummularis* (Plates 17
and 18).

When there is a marked tendency for the lesions **to be-**
come confluent and to melt into each other the result is **an**
eruption of various forms, concave and convex lines—
psoriasis figurata.

Further along in the course of the disease, or in some
cases appearing early in its course, the central part of the
patches tends to disappear and **the** disease spreads peripherally, producing rings of various sizes—*psoriasis an-
nularis, psoriasis circinata* **(Plate** 19).   In this variety,
if **the** circles run together, **the** meeting borders melt
away and give rise to irregular, tortuous scaly bands—
psoriasis gyrata (Plate 20).

Finally, if the lesions grow to considerable size and
new outbreaks occur, large confluent areas result, infiltrated **and** scaly, along with the ordinary, scattered,
variously **sized** lesions—*psoriasis diffusa universalis.*

Psoriasis involves especially the outer skin.   Frequently the process involves the nails, which may **be-**
come milky, break easily, crack, and from time to time
may crumble or be cast off entire.   In rare instances, and
then only after long-continued psoriasis of the hairy **scalp,**
the hair **may** fall out to some extent.

The pathologic basis of psoriasis is an inflammation in
the papillary **layer.**   The vessels are hyperemic, the upper
corium and stratum papillaris are infiltrated **with** serum,
and around the vessels are seen cell-collections.   **The rete**
is somewhat relaxed and edematous ; to this **is** due the
fact that in **the** matured lesions **the** accumulated and

horny epidermis may be rubbed **off by light** scratching, and from the underlying hyperemic blood-vessels bleeding **is readily** produced.

To the rapid reproduction of the epidermis-cells is **due the** fact that the scales are silvery and shining, not **having** time to **go** through the process of cornification. **Old** patches, **on** the contrary, show thickening of the skin due **to** hyperplasia **in** the papillæ and connective tissue, even to the extent of becoming somewhat wart-like (Kaposi).

As a rule, patients are apt to have psoriatic patches **on the** extensor surfaces of the knees and elbows for years, **and** then for **some** unknown reason efflorescences show themselves **on** the **trunk** and limbs.

The scalp and **the** bordering forehead rarely remain **free.** The eruption seldom shows itself on the face, and still **less** frequently on the palms and soles.

In rapidly developed **cases** the inflammatory characters **may** sometimes, **in** the **course** of three or four weeks, show considerable retrogression ; **the** plaques become much flattened and the scaliness less marked. In most cases, either through spontaneous retrogressive changes or **as** the effect **of** treatment, most of **the** patches disappear ; but frequently there remain slightly perceptible remnants on the places of predilection—the knees and **elbows.** When this is the case the patient can be almost sure that the disease **will in some** months begin to spread afresh.

As to subjective symptoms, the most common is itching, **especially** associated with disease in those cases in which **are** marked infiltration **and** inflammation of the papillary layer. In severe cases there may **be** gastric disturbance, restless nights, and the appearance of painful fissures about the joints and flexures, so that the patient becomes incapable of getting about, and is obliged to give up work and for **a** time keep to his bed.

Several atypical **forms** of psoriasis have been described which were distinguished either by peculiar appearances **on the skin** itself **or** by complications with joint- or organic diseases. As yet, it is not proved whether these complications have any relationship to the psoriasis, or whether, as is more probable, they are accidental only.

We have repeatedly observed psoriatics with marked disturbances of the general health. The patients feel weak, show atypical temperature-changes, and complain of unrest, sleeplessness, and loss of appetite. The psoriatic patches may be succulent, elevated, covered with dirty-white scales, and surrounded by an inflammatory halo or band several millimeters in width. Gradually the subjective symptoms abate, the psoriatic plaques flatten, and there develops the usual picture of psoriasis, in which the involution-period is apt to be more rapid than usually observed. This peculiar course and condition we do not venture to ascribe to the cause responsible for the psoriasis; but it impresses one as being due to some toxic agent, as in the erythemata.

We have pictured a case of this kind with dollar-sized and larger plaques with inflammatory appearances (Plates 17 and 18). It was also remarkable in that on the hairy scalp was seated a large and hard scaly or crusted plate-formation of a dirty-white color, which we were able to loosen in mass, representing a cast of the parts. Under this the skin was infiltrated, slightly reddened, and covered with recent epidermis.

In other instances we observed eczema in psoriatic patients which partly masked the clinical picture of psoriasis; an unfortunate complication, inasmuch as the patients were troubled with constant itching; and we frequently, on account of the presence of the eczema, could not treat the psoriasis with the ordinary therapeutic methods. These exceptional cases occurred in uric-acid patients, who were constantly troubled with the disease.

In recent years cases of psoriasis complicated with joint-affections have been described. Such a case was observed in Lang's clinic, and described as *psoriasis ostreacca* by Dr. Deutsch in *Wiener klinische Wochen-*

schrift, 1898, No. 6.   This case, owing to its extent
and the intensity of the process, and the peculiar form
of the efflorescences and the associated joint-affection,
is especially interesting.   Gassmann published a similar
case as *psoriasis rupioïdes*, and Grube several such associated with gout and diabetes.

We have (Plates 21, 21, *a*, and 21, *b*) pictured a similar
case, and saw also, on a visit to the City Hospital in
Ragusa, a second case, which in addition to severe joint-
affection presented horn-like, heaped-up, pyramidal scales
and large, dirty-white, mortar-like crusts.   As yet we
are not certain that these cases simply represent a very
intense process; but must believe that uric-acid is the
additional causative element.

Psoriasis is commonly observed in well-developed,
strong individuals in the prime of life, so that one cannot
say that a cachexia or a general disease is the cause.   [To
this statement there are many exceptions.—ED.]

It may, however, be stated that a hereditary disposition to the disease exists, which is shown in the family
of the patient, in various branches, as grandparents,
parents, or brothers and sisters, although by no means
observed with that regularity and frequency which obtain with hereditary syphilis.

The **prognosis** in psoriasis is usually favorable.
Spontaneous involution of the psoriatic patches of the
first attack and the recurrences is not uncommon.   Besides, modern methods of treatment have an influence.
Severe, complicated cases, such as those mentioned, or
the accidental infection through the fissures which may
occur about the joints of the erysipelas or phlegmon
microörganism, may threaten the life of the patient.

Treatment.—Internal Remedies.—1. *Arsenic.*—(*a*)
In the Form of Fowler's Solution.—Six drops are given
daily, divided into three doses and taken diluted.   Every
day the amount is increased by one drop.   One can in
this manner increase the quantity up to thirty drops
daily, remaining at the dose at which involution of the

psoriasis is observed to take place ; not **to be** discontinued
suddenly when apparently completely **cured, but return-**
ing gradually **to** a less and **less** quantity **(Kaposi).**

(*b*) *As Asiatic Pills.*—

> ℞ Arsenici albi, 0.75 (gr. xj) ;
> Pulv. piperis nigri, 6 (ʒiss) ;
> Pulv. acaciæ, 1.5 (gr. xx) ;
> Pulv. althææ, 2 (gr. xxx) ;
> Aquæ fontan., q. s. ut. ft. pil. No. 100.
>
> Sig.—Three pills to be taken daily.

Every fourth day the dose is increased **by one pill up to**
ten or twelve pills daily ; and in the same manner as **with**
Fowler's solution, gradually lessening this quantity after
an apparent cure. The pills are to be taken immediately before meals. [Less apt to disturb if taken after meals.—ED.]

In addition to these methods of administering this
remedy, **it** may be given by subcutaneous injection—of
Fowler's solution, 0.2 (𝔪 iij) pro die ; of arseniate **of**
sodium, 0.02 (gr. $\frac{3}{10}$) pro die. According **to Ziemssen,**
the official solution **of** arsenite **of** potassium is inappro-
priate **for subcutaneous** injection, owing **to** the method of
its preparation and also to the presence of a fungus which
develops **in it.** He recommends the following : One
part of arsenious acid is boiled with five parts **of** soda
solution till complete solution is effected ; **it is** then
diluted to make one hundred parts and filtered. For
use, put some of the solution in a small tube, which is
stopped with a **wad of** cotton, and it **is** then sterilized **with**
steam. Of this 1 per cent. solution of arseniate of sodium
the beginning injection is 0.25 (𝔪iv) once daily ; after **a**
few days twice daily, and then the quantity of each injection is gradually increased and is administered twice daily.

Danlos and Rille recommend sodium cacodylate for subcutaneous injection (sodii cacodylat., 4 (ʒj) ; aquæ destill.,
10 (fʒiiss) ; daily **a** syringeful.

Herxheimer injects 0.001 (gr. $\frac{1}{60}$) arsenious acid (in
solution) in a skin-vein of **the elbow or** knee region.

Every day the dose is increased 0.001 (gr. $\frac{1}{65}$) till it reaches 15 mg. (gr. $\frac{9}{40}$), at which it is kept till complete disappearance of the efflorescences.

2. *Potassium iodid* (Greve, Haslund) in increasing dosage, beginning with 3 to 4 grams (gr. xlv–gr. lx) pro die, increasing every third day about 1 to 2 grams (gr. xv –gr. xxx), and may even be increased to 60 to 70 grams (ʒxv–ʒxviiss) pro die. Generally, this energetic treatment is well borne, but the large doses should be given while the patient is under direct observation ; the result in many cases is not to be doubted.

3. *Thyroid preparations* (Byrom Bramwell) ; especially of these, however, the more reliable preparation, iodo-thyrin Baumann (Paschkis and Grosz). One begins with 0.5 (gr. viiss) of the commercial triturate, and increases the dose every three or four days by about this same quantity. Untoward heart-action and psychical symptoms are to be guarded against. Should head-pain and heart-palpitation appear, the dose is to be intermitted ; if no symptoms appear, one may increase the dose to 5 to 6 grams (gr. lxxv–ʒiss) pro die. The effect in some cases is surprisingly favorable.

External Treatment.—First of all, softening preparations, as salves, oils, sapo viridis, besides baths and rubber clothing, are employed in getting rid of the scaliness. Only after the scales have been removed is it advisable to begin with those special remedies which are commonly used in this disease. As such, may be named :

1. *Tar preparations :* Ol. cadini (oil of cade), ol. rusci (oil of birch), ol. fagi (oil of beech), pix liquida, ol. lithanthracis (coal-tar), tinctura lithanthracis Leistikow (ol. lithanthracis, 30 (ʒj) ; spiritus, 95 per cent., 20 (ʒv) ; æther. sulph., 10 (ʒiiss)), solutio lithanthracis Sack (ol. lithanthracis, 10 (ʒiiss) ; benzol, 20 (ʒv) ; aceton, 77 (ʒiiss)), liquor anthracis simplex, liquor anthracis compositus (Fischel), liquor carbonis detergens (Wright, Jaddassohn).

These may be applied to the psoriatic areas, either as a

liquid preparation, painting on, or rubbing in, with a brush, as, for example:

R Olei rusci,
Ol. olivæ,   *āā* 24 (f℥vj).

Or in salve form:

R Pix liquidæ,
Lanolini,   *āā* 50 (℥iss).—**M.**
Ft. unguentum.

R Ol. rusci,   20 (f℥v);
Saponis **viridis,**   5 (gr. lxxv).
Lanolini,   75 (℥ij ℥ij)—M.
Ft. **unguentum.**

Of special value is a 10 **to** 20 per cent. tar-salve **with** unguentum caseini, with the addition of sapo viridis (one part **of** sapo viridis to four parts of tar). Finally, tar is **sometimes** used in the form of the so-called tar-baths.

2. *Chrysarobin,* in salve form, 5 to 15 per cent. strength, or **with a** drying vehicle (traumaticin, collodium, linimentum exsiccans, filmogen), **as, for example:**

R Chrysarobin,   10 (℥iiss);
Traumaticin (liq. gutta-perchæ), 90 (f℥xxiiss).

Also as chrysarobin plaster (Beiersdorf), **and a 30 per** cent. collætinum chrysarobini (Turinsky).

With the chrysarobin **treatment the** affected **parts become** white and the **surrounding skin** violet **to** brown. During the application of this remedy and for some days afterward baths should be prohibited, as such may tend **to** bring about a slight **universal** dermatitis.

Lately, Kromayer has recommended chrysarobin-tri-**acetate** (eurobin) and chrysarobin tetracetate (lenirobin).

R Eurobin,   2 (gr. **xxx**);
Engallol,   10 (℥iiss);
Aceton,   10 (℥iiss).—M.
Sig.—External use.

℞ Lenirobin,   5–20 (gr. lxxv–ʒv) ;
  Pasta zinci oxidi,   āā 100 (ℨiij).—M.

Sig.—External use.

℞ Lenirobin,
  Eugallol,   āā 6–10 (gr. lxxv–ʒiiss) ;
  Chloroformi,   50 (f ʒxiiss).—M.

Less valuable is anthrarobin.

3. *Pyrogallic Acid.*—Its method of application is the
same as with chrysarobin ; the urine is to be watched, as
absorption may take place when used too extensively.

Unna recommends " pyrogallolum oxidatum " as a safer
preparation :

℞ Pyrogalloli oxidat.,   5 (gr. lxxv) ;
  Vaselini,
  Adipis lanæ,   āā 25 (ʒviss).—M.
Ft. unguentum.

℞ Pyrogalloli oxidat.,   5 (gr. lxxv) ;
  Vitella recentia ovorum duorum misce intime.

Sig.—To be painted on.

Kromayer applies pyrogallol triacetate (lenigallol) and
pyrogallol monoacetate (eugallol).

℞ Eugallol,
  Aceton,   āā 10 (ʒiiss).

℞ Lenigallol,   1–5 (gr. xv–gr. lxxv) ;
  Pasta zinci oxidi, q. s. ad 100 (ℨiij).

℞ Lenigallol,
  Pasta zinci oxidi,   āā 10–30 (ʒiiss–ℨj) ;
  Vaselini,   q. s. ad 100 (ℨiij).—M.
Ft. unguentum.

Less valuable appears to be the application of gallanol
(Cazeneuve and Rollet), and likewise gallacetophenon.

4. *Sulphur.*—This in the form of the natural spring-water baths; as Vleminckx's solution (liquor calcii sulphurat.). In using the latter the patient is first thoroughly washed with soap and water; immediately thereafter the affected areas painted with it, and **the** patient then gets into a warm bath and remains one to two hours.

Very efficient is the treatment with unguentum Wilkinsoni:

R̥ Sulphur sublimat,
   Ol. fagi,   *āā* 50 (ʒxij);
   Saponis viridis,
   Adipis,   *āā* 100 (ʒiij);
   Cretæ albæ,   10 (ʒiiss).—M.
 Ft. unguentum.

This salve is to be **rubbed** into the affected spots **twice** daily. After a week exfoliation of the epidermis begins; after its completion **a** bath is ordered.

PITYRIASIS RUBRA (HEBRA).

This extremely **rare** disease attacks in the beginning the flexures of the limbs, and may for years be limited to those regions. The affected area is vividly red and is the seat of **a thin,** leaf-like epidermal exfoliation. The patient feels moderate itching, so that the complete picture resembles that of a squamous eczema.

Sometimes, however, the disease spreads over the face **and the** rest of the body. Thickening of the skin does not ensue, but there is **a** condition of hyperemia and scaliness without any further changes. The patients experience, in addition to the itching, a feeling of coldness.

The skin loses its elasticity and is tense and drawn, **so** that ectropion **of the** eyelids and hindrances to movability of the lips and extremities result.

Gradually the hyperemia disappears, and the skin becomes atrophic, paper-thin, and translucent. The sebaceous and sweat-gland ducts atrophy, the hair falls out, and the nails become **fragile.** The skin is easily injured, fissures **occurring** frequently about the joints.

We saw in a case in the atrophic stage, at the end of an intercurrent internal affection, bedsores arise in spite of the most careful nursing.

The **treatment** of this disease is wholly without re-sult. It may be mentioned that Kaposi observed, in a recent case, **cure follow the internal** administration of car-bolic acid.

PITYRIASIS RUBRA PILARIS (DEVERGIE).

Under this name Devergie described some time ago a peculiar disease; since then, C. Boeck, Besnier, and many others have reported similar cases and have added the weight of their opinion to that of Devergie. On the other hand, Kaposi contends that pityriasis rubra pilaris is not a disease *sui generis,* but is identical with lichen ruber acuminatus. Since we do not yet know the features and behavior and external differential characteristics of these two diseases—and indeed their external symptoms are very similar—the question still remains an open one.

The above-named writers base their opinion upon cer-tain clinical points which distinguish this disease from lichen ruber acuminatus:

The appearance of whitish-gray or reddish papules, which consist of hardened epidermis and project from the follicles.

The extensor surfaces of the hand, fingers, and fore-arms, likewise the face, are in the beginning more fre-quently the seat of the disease than the trunk.

The surface of the skin feels rough and uneven, flatten-ing out in the further course of the disease as the papules become more closely set. When this has occurred the diseased skin, instead of showing pointed papules, is cov-ered with small scales (scalp) or with larger lamellæ (palms and soles). The hairs sometimes break off or fall out, and the nails become longitudinally furrowed and broken; both, however, do not happen in all cases and not in like intensity.

The affected skin is somewhat hyperemic. In the beginning the skin immediately surrounding the papules is reddened; later the redness spreads over large areas. The infiltration of the skin and the hardness to the touch do not reach a marked degree.

The subjective symptoms, as itching and sensitiveness to pressure and touch, vary somewhat in different cases. The disease occurs in earlier life, spreads only slowly over large areas or the entire body, and disappears sometimes spontaneously or upon the administration of arsenic. It recurs, however, but never leads to a fatal termination.

LICHEN RUBER.

According to the external appearances presented by this disease, two forms are recognized: *Lichen ruber acuminatus* and *lichen ruber planus.* It must be acknowledged, however, that opinion is divided as to the identity of these processes. [Many writers now consider these two forms as different diseases; and, as already observed, some observers look upon lichen ruber acuminatus as identical with pityriasis rubra pilaris (Devergie).—ED.]

Lichen ruber acuminatus appears in the form of millet-seed, reddish, irregularly-scattered papules which terminate in hardened, horny epidermic points. The papules increase rapidly in number, and form either lines or bands, or cover, in a period of two or three months, large plaques of skin; they are especially thickly set, and contiguous in closely-arranged lines or in large crowded areas, on the flexor surfaces of the extremities, especially the upper.

The skin of the affected areas is then uniformly red, thickened, and crackled. The surface is uneven, furrowed, feels dry, rough, and to the hand passing over it not unlike the surface of a nutmeg-grater. The crowding together of the papules in rows, with linear depressions or furrows between, gives it the appearance of shagreen leather, to which Hebra has aptly likened it.

7

The hairs become atrophic and fall out. The nails lose their brilliancy and become fragile. The palms and soles are the seats of markedly thickened, hardened epidermic accumulations, by which the movability of the hands and fingers is compromised.

The patients, who from the beginning of the disease are troubled with severe itching night and day, grow very nervous and tend to become emaciated or of impaired nutrition. The first-described cases by Hebra ended fatally; but after the adoption of arsenical treatment which he introduced subsequent cases were cured, leaving behind atrophic lines and slightly-depressed furrows.

Lichen **ruber** *planus,* or *lichen planus* (Plates 22 and 22, *a*), as it is usually termed, occurs much more frequently than lichen ruber acuminatus. In this variety the papules appear as millet-seed- to hemp-seed-sized, and are elevated, flat, and waxy. At first limited to single regions, later the papules are found extending over larger areas or possibly over the entire surface. In the center of each lesion is a slight depression or umbilication. The earlier scattered lesions, by new accessions, gradually form band-like, linear, or dime- to dollar-sized, more elevated, dark-red plaques. Most lesions show firmly adherent whitish scales. The increase of the papules and the spread of the disease are seldom so rapid as with lichen ruber acuminatus. The groups remain longer stationary. The involution of the papules begins i.i the middle, the center of the patch or plaque becoming brownish in color, while on the border fresh bright-red lesions continue to appear.

The substratum of the process consists of an inflammatory infiltration in the corium and papillary layer, which leads to the above changes in the epidermis.

According to the degree of hyperemia, and sometimes also to increase in the exudation, depend the clinical appearances. Whether, however, such varying conditions are ever sufficiently marked to influence or change completely the ordinary picture of lichen is very questionable; at all events the appearance of vesicles, for example, as has exceptionally been reported, is not a part of this disease. It is probable, and as Lassar rightly says, that such unusual manifestations are accidental and due to the arsenic administered.

In the beginning the lesions are millet-seed-sized; but they may become hemp-seed- or even pea-sized, and, according to their grouping, may present various pictures upon the skin. For instance, we may meet with diffused, red, slightly-scaly patches on the extremities, and near by or on the trunk scattered papules. Sometimes the lesions form in bands or branches or garland-like rows (*lichen moniliformis*), arranged apparently along nerve-tracts.

On the palms and soles the disease causes thickening of the epidermis (tylosis palmaris et plantaris), and gives rise to the consequences of such accumulations—fissuring, loss of movability, etc.

The mucous membrane of the cheeks and tongue may share in the process. We meet with such as epithelial accumulations in the form of white, irregularly shaped plaques with red, hyperemic edges. Owing to the possibility of mistaking it with other processes—syphilis, for example—it is to be remembered that the disease may also appear on the genitalia. The dark-brown pigmentation, surrounded by fresh papules, the troublesome itching, and the duration of the process, sufficiently characterize lichen.

The disease appears in the adult, mostly in well-nourished individuals. It is neither inherited nor contagious. Other skin-diseases, as, for example, eczema, may occur at the same time, and are sometimes produced indirectly by the lichen, by the attempts to gain relief from the itching by rubbing and scratching. The course is protracted, but not so active or tempestuous as in lichen ruber acuminatus. In how far both processes differ from one another, we are not in position to say. Histologic investigations give no conclusion, the slight differences found are not sufficiently characteristic, and, moreover, we have observed cases in which both forms existed alongside of each other.

The very troublesome itching gives rise to various discomforts, as unrest by day and loss of sleep by night; the appetite is lessened, and when no relief is obtained the nutrition suffers. The patients lose their power of resistance and frequently become the subjects of intercurrent disease.

The **diagnosis** of lichen ruber is, if a careful consideration of the above-described symptoms is given, and no other skin-disease temporarily masks the symptoms, not difficult.

Many forms of psoriasis, especially when accompanied by itching, may occasionally give rise to some confusion in the diagnosis. The more frequent occurrence, the greater participation of the extensor surfaces of the elbows and knees, the less infiltration of the skin, and the loosely adherent silvery-white scales, speak for psoriasis.

Eczema squamosum will usually yield a history of pre-existing vesicles, and eventually in its course fresh outbreaks of similar lesions point to this disease. Pityriasis rubra (Hebra) is distinguished from lichen ruber by the absence of infiltration, and also by the thin atrophic skin.

The so-called psoriasis syphiliticus—papulosquamous syphiloderm—and the mucous patches (resembling somewhat the mouth-patches of lichen) of syphilis are associated with other characteristic symptoms of this disease. The mucous patches have not the characteristic red edge of lichen-ruber plaques. Orbicular papules of a syphilitic character about the genitalia, which bear resemblance to those of lichen ruber planus, are usually found with a history of syphilis and other symptoms of that disorder, such as plaques on the mucous membranes, hair-loss, glandular swellings, etc. In addition these syphilitic papules are seldom dry as are those of lichen ruber.

Treatment.—The itching is to be treated by local douches, baths, and alcoholic lotions of carbolic acid, salicylic acid, menthol, etc. Lassar touches the efflores-

cences with the galvanocautery. To promote involution
of the lesions Unna advises :

℞ Ungt. zinci benzoimat.,   30 (ʒj) ;
   Acidi carbolici,   1.25 (gr. xx);
   Hydrargyri   chlorid.
     corros.,   0.03–0.3 (gr. ss–gr. v).—**M.**
Ft. unguentum.

Or

℞ Acidi carbolici,   5–10 (gr. lxxv–ʒiiss) ;
   Hydrargyri chlorid.
     corros.,   1–5 (gr. xv–gr. lxxv) ;
   Creosoti,   2 (gr. xxx) ;
   Collodii,   50 (fʒxiiss).—**M.**
Sig.—Apply with a brush. Use with caution.

Arsenic internally, as in lichen ruber acuminatus, is
also valuable.

The other recommended remedies, as potassium chlorate
(W. Boeck), asafetida and mercurials (**T.** Fox), do not seem
to have any appreciable influence upon the disease.

LICHEN SCROFULOSORUM.

This disease is met with in young individuals, especially between the ages of fourteen and twenty. The skin
of the extremities or the trunk is beset with grayish rough
papules, which occur in bunches or scattered over large
surfaces. It is without subjective symptoms, and the
patients, therefore, often carry the rough patches until
their attention is called to them accidentally. The
affected skin is rough and greasy to the touch, and in
places almost smeary ; never so dry as in ichthyosis and
chronic eczema. The seat of the papules is in the follicles and the perifollicular tissue. The epidermic plugs
protruding from the follicles of the sebaceous glands
often contain a hair or are covered with small thin scales
which may be easily brushed off. Not seldom, especially

in badly-nourished and run-down individuals, the papules are brownish from the admixture of blood-pigment ; often, however, particularly on the lower extremities, they may be bluish- or even brownish-red (*lichen lividus*).

In some cases inflammatory action occurs and acne-pustules result (acne cachecticorum).

As already remarked, this disease is met with in pale and badly-nourished young persons. They not uncommonly show so-called scrofulous swellings in the glands of the neck, or may show fistulous purulent tracts. Impoverished circumstances and lack of care lead to other diseases of the skin, as, for example, to eczema around about the suppurating glands or about the genitalia ; further, to pustules, ecthyma, and furuncles, which, however, are to be considered accidental, and not necessarily part of the symptoms of lichen scrofulosorum.

Treatment.—The chief consideration is the general treatment, which has for its object improving the nutrition with tonics, such as iron, arsenic, cod-liver oil with phosphorus, iodin, change of scene, etc. Locally, according to the practice of Vidal and Hebra, the affected areas are rubbed with cod-liver oil.

KERATOSIS PILARIS (LICHEN PILARIS).

This is a disease (Plate 38) frequently seen upon the extensor surfaces of the lower and upper extremities, characterized by the presence of pale-red papules surmounted with epidermic scales. After removal of the epidermic scales a rolled-up lanugo-hair is observed. The parts have a goose-flesh appearance, and may be considered physiologic in the period of puberty [?—ED.] ; similar papules are also seen in ichthyosis. In this latter disease the condition may be more or less universal.

ECZEMA.

This widespread and therefore important disease is an inflammatory affection of the skin, accompanied by the

subjective symptoms of itching **and** burning. **Many dif-**
ferent clinical pictures are presented in eczema, **so much**
so that formerly these several varieties **or** manifestations
were considered different diseases; Hebra proved, **how-**
ever, that **the** various phases and clinical pictures really
expressed but one disease. The disease may be acute or
chronic.

Acute eczema begins with the appearance of irregularly scattered red papules (*eczema papulosum*), which
give rise to troublesome itching. The papules may retrogress, the redness disappearing **and** a superficial epidermal
exfoliation taking place. Frequently, however, through
intensity of the inflammatory process, these lesions change
rapidly into vesicles (*eczema vesiculosum*).

If the intensity of **the** process continues, there **arise**
numerous millet-seed-sized **to** lentil-sized vesicles **and**
small blebs (the latter rarely). In the beginning **or earli-**
est stages these lesions have serous contents, which soon,
from the admixture **of** cell-elements, become milky and
even purulent (*eczema pustulosum*). The overlying epidermis is either broken by scratching or is rubbed off, and
the red surface exudes a liquid secretion.

Sometimes **the** lesions dry to yellowish crusts, which
when mixed **with** blood, which sometimes exudes from
the hyperemic rete or results from scratching, give **rise**
to brownish or even blackish crusts (*eczema crustosum*).
Very rarely, and then only **as a** consequence **of** violent
scratching, is any loss of substance noticed beneath the
crusts, so that when the process has run its **course** and
healing has taken place by a regeneration of the epidermis, no scarring remains.

Frequently, also, acute eczema appears as a diffused
redness and swelling (*eczema erythematosum*). In many
of these cases, on passing the finger over such affected
areas, **one may be** able to detect slight, scarcely perceptible elevations **or** irregularities, from which vesicles may
develop.

The patient first feels **a** sensation of tenseness in the

affected areas, which soon changes to intense burning and itching. The vesicles become **confluent, new** outbreaks rapidly **taking place ;** the part **is soon deprived** of its epiderm, and there appears **a reddened, oozing** surface, the **base** of which consists of the rete **Malpighii** and papillary layer. The profuse **secretion mixes with** the epidermic cast-off cells and **becomes** thereby thicker and more smeary (*eczema madidans, eczema rubrum*). If the affected areas are not confluent, **or** if **the** intensity of the process and the consequent secretion subside, the parts become covered with extensive yellowish translucent lamellæ, which crack, and through such fissures underlying collected liquid oozes **out** (Plates **23** and 23, *a*).

If the hyperemia, and with it **the** swelling, subsides, the secretion likewise correspondingly lessens ; the epidermis begins to re-form, and the epidermic cells lie upon the still reddened, **infiltrated** skin as loosely attached scales (*eczema squamosum*). This scaly condition **may** persist for some time **or** rapidly disappear, and **a normal** condition **be** re-established.

As already stated, all stages of **acute** eczema may pass **directly and** rapidly **to** cure. More frequently, however, **we** observe that the papular or vesicular stages change into the squamous **stage** or variety. Often we meet with a squamous type on **one** part of the body, on another a crusted form ; this is especially noticeable in universal eczema and **in** recurrent or relapsing forms.

As **a** peculiarity of eczema, it may be mentioned that often a long-continued mild eczema, to which the patient gives **but** little thought, without recognizable cause develops **into acute** eczema on distant situations. Many authors (Kaposi) look upon such as due to vasomotor neurosis ; but this itself must have a foundation. Many individuals have at certain seasons of the year an unmistakable disposition to eczema, and even after freedom for **a** number **of** years the old trouble returns.

Acute eczema, fortunately, is **rarely** encountered as a generalized disease ; but it produces **a** severe, sometimes dangerous condition when it involves the whole surface in various degrees of severity. Some parts of the body, as the face, the genitalia, and the hands, are markedly swollen, and the patients experience tension, burning, and itching, which, with the accompanying fever and systemic disturbance, are very troublesome. The clothing adheres to the oozing places and causes further irritation; the patients find no relief or rest and lose sleep. They complain of weakness, loss of appetite, and frequently chilliness; and these conditions, together with imperfect nourishment and by loss of the blood-plasma, may lead to a grave issue. [Such extreme cases must, however, be rare, and it is even questionable in those instances whether the disease is not complicated or other than eczematous.—ED.]

The duration of universal eczema is uncertain, since after subsidence of the acute stage it only partly disappears, remaining on several parts as chronic eczema.

Of the localized forms of acute eczema, the most frequent is eczema of the hands, these parts being the most exposed to external irritating agencies. It appears with swelling of the back of the hand and fingers, which sometimes extends up the forearm. The hard and thick epidermis of the palms is slowly cast off. Frequently painful fissures (rhagades) arise, and sometimes the surface around the nails becomes raw looking, with at times granulation-tissue formation, so that for a considerable time the patient is unfitted for using the hands. The same appearances and conditions obtain with acute eczema about the feet, only on these parts the disease is much less common.

The face is a frequent site for acute eczema (Plate 24). Marked swelling of the eyelids, cheeks, nose, lips, and even the ears is noted, and gives rise to a feeling of tenseness. Not infrequently eczema of this part is mistaken for erysipelas faciei. This latter, however, is wanting in papules, vesicles, and pustules, and consists of a diffused firm infiltration, usually with sharply-defined borders, with tenderness and continued high fever. It is, unfortunately, seldom that the eruption on all parts in acute eczema of the face so completely disappears that there is but slight prospect of recurrence or relapse; the simultaneous involvement of the ear-lobes with the face is especially unfavorable for such outlook. An uncomfortable result or consequence of acute eczema is the dryness and brittleness of the skin, which in spite of apparent cure remain and give rise apparently to recurrence.

Acute eczema of the genitalia occurs more frequently in men, and is accompanied by great edema and swelling of the penis and scrotum. It begins with a feeling of weight and tenseness, and obliges the patient to seek rest in the recumbent posture. Soon the skin of the affected parts becomes inflamed and fissured; there is also abundant oozing, which adds to the patient's discomfort, inasmuch as crusts form which crack or are more or less torn by the scratching and rubbing and cause painful burning. In women the disease usually first affects the labia, and then rapidly involves the genitocrural folds, and sometimes spreads down the thighs.

Eczema intertrigo is not uncommon, and may involve considerable surface; it is accompanied with a scanty secretion and with constant casting off of the epidermic cells, which together constitute a greasy covering over the reddened corium. The process is most frequently observed on contiguous surfaces, as the anal fold, under the breasts, in the flexures of the legs and arms, and in many other regions in fat children and corpulent adults.

Chronic Eczema.—Morphologically chronic eczema is but slightly different from acute eczema. Clinically, however, there are many points of difference in the course of the affection which distinguish the chronic process from the acute. Chronic eczema arises either in the wake of a rapid incomplete involution of the acute disease, as already stated, or an acute eczema gradually becomes less and less marked and passes almost imperceptibly into the chronic process.

The chief forms of chronic eczema are the oozing (*eczema madidans, eczema rubrum*) and the scaly types. Although sometimes papules and vesicles of a markedly inflammatory character may be noted from time to time, the chronic type is characterized essentially by persistence, frequent recurrences, obstinacy, and rebelliousness. To these characteristics may also be added consecutive changes which are brought about by the chronic disease: Brittleness and vulnerability of the skin, disposition to branny scaliness, scurfiness, and finally the painful fissures which usually appear in the flexures and about the joints. As a further result of the chronic disease may be mentioned an increase in the pigmentation of the affected regions, sometimes thickening of the epidermis, thickening of the corium, and increased connective-tissue growth. These latter may under certain circumstances, especially when involving the lower leg, almost approach an elephantiasic condition in appearance.

Among the subjective symptoms stands, first of all, the intense itching, which is the source of so great distress to patients that they continually rub and scratch, both when clad and unclad.

It is rare that chronic eczema involves the entire surface; as a rule, only certain parts are predisposed to it. There are several places of predilection:

Chronic pustular eczema of the scalp, frequently associated with eczema of the ear-muscles and the face. The scalp is covered with broken-up yellowish or yellowish-green, frequently brownish crusts. Here and there in the hair are found cast-off or rubbed-off fragments of crusts, and in some cases also lice and nits. On removing the crusts from the underlying skin the latter is seen to be red, oozing, and deprived of its epidermal covering. The hairs become matted or project irregularly through the crusts. This condition is not infrequently seen in women and children as a result of pediculosis capitis. These parasites may be primary (the eczema resulting) or they may be secondary. The children have, moreover, frequently swelling of the cervical glands, which the mother is apt to look upon as scrofulous. If this condition of pediculosis is neglected, and to it added extraneous dirt and filth, the hairs become tangled in masses or into long, thin bunches (*plica Polonica*).

Chronic eczema of the face seldom involves this whole region; usually only certain parts, such as the mouth, lips, ears, eyebrows, and eyelids.

A special variety of eczema of the face is observed in infants, in which the face and ears are covered with crusts (*crusta lactea*). The ears, cheeks, and brow are most commonly the seat of this troublesome and itchy affection.

Eczema of the lips, which often occurs in association with eczema of the nose, leads to thickening of the border, and often of the entire lip, with fissuring of the vermilion; even after complete healing of the lesions the lips may remain permanently enlarged, with linear cicatricial or atrophic furrows.

Eczema of the genitalia and anal furrow leads to many consequences, brought about by the itching and scratching: Thickening of the skin, growth of the chronically inflamed furrows, etc.

It remains to mention eczema of the flexures of the extremities, of the nipples, of the mammæ, and of the navel, which presents symptoms in no respect different from the disease in other parts.

The occupation of many individuals provokes *eczema of the hands,* fingers, and even the finger-nails (trade-eczemas). These eruptions are not only characterized by vesicles and pustules, but the epidermis of the palms and of the fingers is thickened, brittle, and fissured, so that the many places deprived of their epidermis render it painful for the patient to work. A similar condition of affairs, in somewhat less degree, occurs also on the feet.

We have yet to refer to certain eczematous eruptions known as *impetigo faciei contagiosa* or *parasitaria.* Blebs, crusts, and scab-formations, either in large confluent areas or in groups, are noted, which may be surrounded with scattered red follicular elevations. Although it is not yet positively proved, nor the causative fungus found, yet the apparent spread of the disease from one to another—its easy auto-inoculation—makes its contagiousness probable. We have seen such eruptions in young persons, occurring in patches and groups on the face and neck, on the breast, and even on the forearm (Plates 27, 28, and 28, *a*). [Many of these cases (not those pictured) are considered by numerous writers as examples of a distinct disease—impetigo contagiosa.—ED.]

In conclusion, we will make mention of *eczema marginatum* (Hebra), as a special form of eczema. It appears in palm-sized areas, confluent circles, and ellipses, which show vesicles on their borders ; the central parts being either covered with scabs and scales, or, if of long duration, showing a somewhat dark pigmented skin. The sites chiefly affected are the inner thighs and the genitalia. A variety in its beginning or early stage is shown in Plate 26. [This is again referred to under the head of ring-worm.—ED.]

The so-called *eczema seborrhoicum* Unna has considered a disease *sui generis*. This develops, as a rule, from a slight and unnoticed seborrhea of the hairy scalp. Marked aggravations, such as hair-loss, increased collection of scales and crusts, intense itching or oozing, lead the patient to seek medical aid. From the scalp proper the disease spreads to the forehead and temples, with a sharply-defined border which is sometimes quite red and is covered with yellowish, greasy scales ; the disease not infrequently also invades the ears and neck. Unna recognizes three varieties of eczema seborrhoicum—the scaly, the crusted, and the oozing.

In addition to the regions already named as the common sites of this manifestation (eczema seborrhoicum), the disease may also attack independently, or, more commonly, conjointly, the sternal region, where roundish or oval patches [seborrhœa corporis of Duhring.—ED.], finger-

nail in size, appear singly or in groups, each spot of yellowish color with a narrow red border; the axillæ, where
the affection may present a red, serpiginous, advancing
line; the flexures of the upper extremities, the dorsal
surfaces of the hands, the buttocks, the hips, the anal
region, and the genitocrural folds.

The affection is to be differentiated from other eczemas
and from psoriasis. Of special value in the differentiation
are the spread from above down, the history of a pre-
existing seborrheic affection of the scalp, and the peculiar
appearances of the individual patches or efflorescences.

Therapeutically, Unna recommends especially sulphur,
in combination with zinc oxid in the oozing form; for the
crusted and scaly varieties chrysarobin, pyrogallol, and
resorcin have proved of value. [These several remedies
are employed in the manner advised under seborrhea and
psoriasis, but **weaker**. Chrysarobin, if employed, should
be used cautiously.—Ed.] Internal medication in this
disease seems without influence.

Diagnosis of Eczema.—When the symptoms are
considered, it will be seen that **acute** eczema is scarcely to
be confounded with any other skin-disease; at the most,
the acute face-eczema with erysipelas already mentioned,
the differential points of which have been pointed out.

Chronic eczema, on the contrary, may, when of long
duration and from its tendency to scaliness, be confounded
with psoriasis and with lichen ruber planus. It is to be
remembered that chronic eczema often has its beginning in
the acute type—that is, there is an entirely different history from that of the other diseases named; and that on
one or more regions outbreaks of an acute character may
occur from time to time which are quite diagnostic.
Eczema is, moreover, chiefly an affection of the epidermis
and rete, and is distinguished from psoriasis in that it
does not appear in numerous, uniform plaques as does
the latter. In lichen planus the papules arise from infiltration of the skin, with less scaliness in disappearing,
and never present an oozing surface. The subjective

symptom—namely, itching—occurs always in eczema, seldom in psoriasis, but frequently, however, in lichen.

Prurigo, ichthyosis, lupus erythematosus, tinea tonsurans and circinata, and favus can scarcely be confounded with eczema. On the other hand, however, a combination of one or several of these diseases with eczema is not a rarity.

The causes are divided into two classes: One comprises those cases in which the disease seems to have been excited by external irritants—external causes; the other, those cases which have been called forth by some general disturbance of the whole organism—internal causes—symptomatic eczema.

By far the more frequent are the first named—mechanical, thermal, and chemic irritation. By eczema due to mechanical irritants we mean those cases brought about by pressure or rubbing, especially if the skin had been previously subjected to heat or irritated in any way. In such instances the constant rubbing of the clothing and the pressure and irritation of bandages suffice to call forth mild forms of the disease. In this connection also should be mentioned those diseases in which itching is a prominent symptom, and necessarily gives rise to rubbing and scratching, and resulting eczema: Lousiness, scabies, prurigo, pruritus cutaneus, urticaria, lichen ruber, ichthyosis, and pemphigus pruriginosus. Among the mechanical causes belong also circulatory sluggishness or congestion due to varicose veins in the lower extremities, especially the lower part of the leg, and sometimes the scrotum. The itching induced by the congestion or blood-stagnation causes the patient to rub and scratch. The epidermis, thinned by frequent hemorrhage or by exudation in the cutis, is easily injured. The repeated eczematous outbreaks give rise to new inflammations and changes; the subcutaneous tissue grows, is thickened; the blood- and lymph-vessels are in part dilated, partly new formation; many anastomoses of these (varicosities) arise anew; the connective tissue immediately surrounding these becomes thickened and increases; with time it becomes still more marked, more or less sclerosed; the affected part increases in volume, and we have the picture of elephantiasis.

Thermal irritation, as, for example, in boiler-makers, often leads to diffused inflammatory disturbances either of the hands, face, or breast-region (*eczema caloricum*), in which there is marked vesicle- and bleb-formation. The heat of the sun (*eczema solare*), as, for example, in rowers and bathers, calls forth, for the most part, papular eczemas.

Frequently we see in long-continued sweating a minute papular or vesicular eruption (eczema sudamen). The profuse sweat-secretion collects either in the ducts of the sweat-glands, lifting up the epidermis, or, also in addition to this, by serous oozing out of the papillary vessels and collecting in the epidermic layer. The rubbing of the clothing or the rubbing and maceration of contiguous surfaces add to the condition and lead, in the further course of the disease, often to true eczema.

Finally, as to the numerous chemic irritants, as, for example, arnica tincture, which is a popular remedy for wounds and injuries; the resins, as turpentine, a constituent of various plasters, and which is also used by many persons in their work, as painters, printers; many medicinal substances, as croton oil, cantharides, mustard, iodoform, sulphur, carbolic acid, corrosive sublimate, old mercurial salves, potash solutions, lye, soaps (owing to the excess of free alkali), particularly in washerwomen; and macerating poultices of cold water, or as a result of cold-water cures (the cutaneous irritations formerly looked upon as "critical" eruptions) (Plates 14, 24, and 25).

The symptomatic eczemas result from various diseases which involve the organism and engender in the skin a state of irritation or vulnerability. It is especially in those general states of the health which bring about depressed nutrition and reduce the individual power of resistance, that the skin is responsive to the slightest irritation.

In this class belong scrofulosis, rachitis, diabetes, gout, excessive corpulence, and the various anemic and dyspeptic conditions which especially dispose the peripheral parts of the body (head, hands) temporarily to eczematous outbreaks.

Course and Prognosis.—Concerning the course of acute eczema there is but little to say. The slight, localized acute forms disappear in two to four weeks. On the contrary, generalized acute eczema terminates for the most part, at least in certain regions, in the chronic form.

The course of the chronic form depends upon the causes which have provoked the disease and upon the changes which have been brought about by it, such as thickening of the skin, fissures, etc. Chronic eczema is not infrequently associated with furunculosis, the latter dependent doubtless upon the scratching and the consequent ready inoculation by the cocci.

As troublesome and obstinate as eczema is, nevertheless one can say, in general, to the patient that recovery is probable. If the cause disappears or is modified, or if the patient avoids the exciting factors, very often slight local therapy will suffice to remove the disease.

Eczema heals without leaving any traces worthy of mention; at the most, here and there some slight pigmentation or insignificant thickening of the skin. As it is *par excellence* a disease of the epidermal layer, no scarring, even in the pustular form of the disease, is left; and should such be observed, is due to accidental causes. Syphilitics, in order to conceal the fact that they have had syphilis, occasionally state that they have suffered from eczema which had been preceded by nerve or organic disease; such a statement, however, is not to be believed if an examination discloses scar-formation occurring in groups and pointing to a pre-existing syphilitic manifestation which had disappeared spontaneously or as the result of treatment.

Internal Treatment.—Especially by the French writers, in all cases of acute and chronic eczema extensive

8

dietetic directions and a number of internal remedies are recommended. Up to the present, however, proof is wanting that all cases are in reality dependent upon constitutional causes, diathesis, etc. ; the probabilities, on the contrary, are rather against such acceptance. The constitutional treatment will therefore be limited to those cases in which there is some disease or functional disturbance of some other organ, as the possibility of some connection between the skin-disease and such may exist. A persistent anemia is to be treated by appropriate remedies; and in cases of diabetes, nephritis, uric-acid diathesis, oxaluria, the proper dietetic directions should be given and alkalies, diuretics, etc. ordered. It must be admitted that better results are to be obtained when attention is also given to the general health than when treatment is directed to the skin alone. In fact, for successful treatment each individual case demands careful study.

External Treatment. —(*a*) **Acute Eczema.** — In eczema intertrigo and papulosum dusting-powders, such as starch, talc, or this combination :

R̥ Amyli oryzæ,   100 (ʒiij) ;
    Zinci oxidi,
    Pulv. iridis **florent.,**   *āā* 5 (gr. lxxv).—**M.**
  Sig.—Dusting-powder.

When the inflammatory symptoms are of high grade ice-cold poultices, aluminum acetate, poultices of 2 per cent. resorcin solutions, 2 to 5 per cent. tumenol solutions (Neisser), and similar applications are to be recommended.

If itching is troublesome, it can be moderated or controlled by applications of alcoholic solutions ($\frac{1}{2}$–2 per cent.) of carbolic acid, salicylic acid, with subsequent powdering, and finally with weak tar-applications. Most authors advise against the application of tar so long as oozing is present; but Lassar, on the contrary, sees no contraindication to its employment in such cases.

In the crusted stage or forms of the disease the soften-

ing salves and oils are especially useful, especially that
sovereign remedy, the unguentum diachyli Hebræ. In
persistent scaly forms salves applied as plasters, such as
vaselin, unguent. aquæ rosæ, unguent. zinci oxidi, ung.
Wilsoni, Lassar's paste, unguent. caseini, with or without
other medication, and cooling salves (Unna) :

R̓ Lanolini,
    Adipis benzoinat.,
    Aquæ rosæ,
Ft. unguentum.

          10 (ʒiiss) ;
          20 (ʒv) ;
          30 (ʒviiss).—**M.**

R̓ Lanolini,
    Zinci oxidi,
    Olei olivæ,   equal parts (Ihle).—**M.**
Ft. unguentum.

R̓ Zinci oxidi,
    Amyli,
    Lanolini,
    Ol. olivæ,   equal parts (Berliner).—**M.**
Ft. unguentum.

(*b*) Chronic **Eczema.**—In addition to the various
local remedies mentioned above are to be commended softening salves, salicylated soap-plasters, and rubber fabric.
In those cases of considerable thickening and epidermic
accumulation in which tar fails to soften and relieve,
strengthen the tar by the addition of sapo viridis (equal
parts) and carbolic acid. Eventually, β-naphthol salve,
pyrogallic-acid salve, chrysarobin salve (1 : 10–1 : 50 vaselin) ; canterizations with caustic potash solutions of varying proportions, 10 to 50 per cent.

PRURIGO.

Prurigo (Plate 29) is a chronic and **extremely troublesome** disease, persisting, by frequent and repeated **recurrences** and relapses or continuously with exacerbations,

throughout life. The accidental **secondary lesions of the skin** are more **conspicuous** than its **own** pathologic **products.** The **disease** begins in childhood, **in** the first or second **year of life,** with outbreaks **of** intensely itchy **hives. The wheals** and scratch-marks may be made to disappear **by** means of baths **and care of** the skin ; but **soon recur.** The wheals **repeatedly make their** appearance, finally resulting in **the formation** of papules. These **are** pin-head in size, pale **or pale red,** and itch intensely, so that they are not infrequently observed **covered** with blood-crusts. Their **sites of** predilection are the **extensor** aspect of **the** lower **leg, the** thighs, the **sacral and gluteal** regions, **and** the extensor surface of the **arms, both upper** and lower parts. These prurigo-papules are scarcely **elevated** above **the level** of the skin ; only **by** persistent rubbing do **they become** prominent. When scratched open they **become** depressed **and a blood-crust** marks the site ; this disappears and **leaves behind a** white scar or speck. The flexures of the knees, **groins, and** elbows, likewise **the face, are** usually uninvolved, **and are** soft, white, and **moist, so long at least as they** remain free **from** eczematous **manifestations, which in** severe cases are **often** associated **or result from** the persistent **irritation and** scratching.

The milder grade of prurigo **is often without** striking **or urgent** subjective annoyances, **in consequence of** which **it** may lack the resulting secondary phenomena. This **type** may, by frequent baths and great care of the skin, **in** individuals favorably circumstanced, be kept stationary and eventually cured. **In** many **such** cases outbreaks are often limited to **the lower** leg and thigh, and **at** the most appear only **in** winter **and** for **a** short time.

These milder types **of** the **disease are** usually designated *prurigo mitis,* in contradistinction **to** the severe forms—*prurigo agria* or **prurigo ferox. In the** latter **variety** of the disease the **outbreaks of** prurigo-papules **are** so numerous and the consequent itching **so** intense that **the** patient is obliged to be constantly **rubbing** and scratching. The skin becomes covered with roundish and linear, brownish, dry blood-crusts, which may be surrounded by an inflamed red or purulent areola. Near by are also to be seen recent red or older white scars. Owing to the repeated cutaneous outbreaks, and the resulting hyperemia and persistent scratching in trying to obtain relief, the skin becomes more or less pigmented, is noted to be hard, rough, and board-like, and can scarcely be lifted in folds. In severe cases the lanugo-hairs are wanting, or here and there are broken off or pulled out by the constant scratching. Especially about the knees and ankle-joints the skin is thickened and shows deep furrows. The intense irritation of the skin, added to by the constant scratching, produces infection and leads to chronic inflammation of the lymphatic glands, more particularly of the femoral, inguinal, and axillary glands. The patients are troubled night and day by the itching, look pale and badly nourished, and are often looked upon by their associates as suspiciously scabietic and are avoided.

This disease disposes the affected individual to eczema, which may attack the few free places in the flexures of the joints and on the face. Besides, pustules and ecthymata on the extremities are not uncommon complications or additions.

The **diagnosis** of prurigo, when the disease is not complicated or masked by a coexisting eczema or scabies, is not difficult, the characteristic symptoms already described, the localities affected, and its course furnishing sufficiently characteristic points : excepting from this statement the earliest stages, when the disease usually presents solely urticarial symptoms.

Etiologically there can be recognized but one positive factor, and that is heredity, inasmuch as it is often observed that several children in one family are affected.

Treatment.—Prurigo-patients are, as a rule, weakly, and are slow in development and ill-nourished, and for these reasons an effort should be made to build up the general health. **Internal** medication (carbolic acid, menthol) is without direct **effect on the skin** ; but as supporting and alterative remedies **may be** mentioned cod-liver oil alone or with **iodin** (iodin, **0.10 (gr. iss)** ; cod-liver oil, **100 (℥ij)), one or two teaspoonfuls, t. d.), and** phosphorus, as in the following :

R	Ol. morrhuæ,	30 (f ℥j) ;
	Phosphori,	0.01 (gr. $\frac{3}{20}$) ;
	Acaciæ,	
	Sacchar. alb.,	$\bar{a}\bar{a}$ 15 (℥ss) ;
	Aq. dest.,	40 (f ℥j ℥ij).

Sig.—One **to four teaspoonsful, t. d.**

Of external applications, tar deserves **most** prominent mention, applied thoroughly ; sulphur (Vleminckx's solution (liquor calc. sulphuratæ), sulphur **salves**) ; Wilkinson's **ointment** (a course of **ten to twelve** rubbings) ; β-naphthol (**5 per cent.** salve **in courses of four** rubbings, and after each course a bath). In addition, sweat-baths (hot baths followed **by** hot pack) ; subcutaneous injections of pilocarpin, 0.01 (gr. $\frac{3}{20}$) each dose ; internally jaborandi-leaves as infusion, 4 : 100 ; and sulphur **baths.**

Murray and Hatschek recommend massage **of the affected** skin, which is said to have **a** remarkably favorable influence upon the itching.

NEUROSES.

In the descriptions of some of the preceding diseases reference was made to the fact that they originated from or were influenced or modified by irritation of the nerves ; diseases which might well **be** termed *trophoneuroses*. It is our purpose, however, to consider here the cases which belong strictly to the neurotic class, in which itching is the essential symptom ; those disturbances of sensibility **which are** not associated with **any external** cause and without **primary** anatomic changes **of the** skin. This may be present in mild **or severe** degree.

Pruritus.—The extreme sensibility or irritability of the skin characterizes itself by itching—*pruritus cutaneus, pruritus.* This affection may occur as *pruritus universalis* or *pruritus localis.* The patient suffers from attacks of violent itching of the skin, so extreme in its intensity that he cannot withstand the desire to rub and scratch, nor usually stop till the skin is reddened or excoriated, and some parts scratched open and bleeding. The itching is usually then replaced by a feeling of burning, and the patient feels weak or exhausted by the effort and the suffering. The attacks are most common in the evening, especially when undressing, and through the night, so that often sleep is broken or fitful. The skin shows diffuse redness, or at the most urticarial wheals near the blood-crusted excoriations; it is frequently found dry, is seldom moist, and after long duration of the disease brownish colored. The sweat-secretion is mostly limited to the joint-flexures. In young individuals disturbances of digestion are noted, and in women disturbances of the sexual organs are often associated with the cutaneous affection. Mental emotions may also have an influence in promoting cutaneous pruritus.

Of troublesome nature is the pruritus of those advanced in years—*pruritus cutaneus senilis*—which may persist to the end of life. Pruritus due to other causes than advanced age may be benefited and relieved; and even the pruritus of senility may often be ameliorated and occasionally temporarily or permanently controlled.

The diagnosis is not always possible upon first sight. One must carefully consider the various dermatoses of which pruritus may be a symptom; also the possible presence of parasites must be excluded.

Treatment.—In the treatment of pruritus the possibility that certain diseases may through noxious influence be causative must be considered: Diabetes, gout, stomach and intestinal disease, liver-affections, and disease of the genito-urinary apparatus in women. If any one of these causes is found to be operative, then the treatment must be directed toward its removal; or, if this is not possible, then toward its modification.

The remedies, both internal and external, which have been recommended for the treatment of this disease are very numerous. We name as the first in importance in the constitutional treatment, sodium salicylate; also, atropin, quinin, pilocarpin, tinctura gelsemii. Externally baths and douches are recommended; in many cases a low temperature of the water, in others a high temperature, seems to be more valuable. Of the external remedies controlling the itching, which may be applied as lotions or salves, are the following: Carbolic acid, salicylic acid, ichthyol, naphthol, tar, chloral hydrate, camphor, menthol, thymol, etc.

The following prescriptions may be given:

℞ Acidi carbolici,   4 (ʒj);
    Aceti aromat.,   200 (f℥vj).

Sig.—Two tablespoonsful to a quart of warm water; to be applied daily, and after it dries on the following powder to be dusted over:

℞ Bismuthi salicylat.,   20 (ʒv);
    Amyli,   80 (℥iiss).

Or the following lotion may be used:

Hydrarg. chloridi corros., 0.03–0.3 (gr. ss–gr. ivss);
Ammonii chlorid.,   0.12–0.5 (gr. ij–gr. viiss)·
Acidi carbolici,   4 (ʒj);
Glycerini,   60 (f℥j ʒvij);
Aquæ rosæ,   120 (f℥iij ʒvj).

Sig.—Apply morning and evening.

Or the following:

℞ Chloral. hydrat.,
    Camphoræ,
    Acidi carbolici,
    Glycerini,   equal parts.

Sig.—Apply morning and evening. Use with caution.

The most frequent pruritus limited to a region is **pruritus pudendorum.** The external genitalia and frequently also the vulvæ (**pruritus vulvæ**) are attacked by intense itching, and the mechanical irritation produced by attempts to gain relief results in thickening, hypertrophy, and catarrhal affections of the mucous membrane.

In men, mostly in those of advancing years, the itching may be limited to the scrotum (**pruritus scroti**) and perineum, and leads quickly to eczema and the above-mentioned changes. Sometimes the urethral orifice, the urethra itself, and the anal crease are also affected. **Pruritus ani** is frequently associated with the various diseases of the rectum, as hemorrhoids, fissures, etc.

Treatment.—In these various local forms of pruritus attention is always to be given to the possibility of its being due to the various diseases named (hemorrhoids, Oxyuris vermicularis, fissures, endometritis, malpositions of the uterus, etc.). The remedies already mentioned in the treatment of general pruritus are also to be advised in the treatment of the local forms.

Anesthesia of the skin we have almost always observed circumscribed in character. It results from some disturbance of the nerve-branches or from disturbance of the central nervous system. Two forms are recognized—one in which the anesthesia is to temperature and the other to the touch. Complete disappearance of the sensibility of extensive areas is sometimes noticed, as, for example, in lepra anæsthetica.

ANOMALIES OF THE EPIDERMIS.

CALLOSITAS.

The epidermis is often produced in excessive quantity, and is cast off in small scales or large lamellæ; or the horny cells remain and result in thickening and callosities. *Callositas* (*tyloma*, *callus*) is a thickening and hardening of the epidermic layers, which may become several

millimeters thick. The form of these thickenings depends
somewhat upon the character and extent of the pressure
which has called them into existence. The sensibility in
the part is more or less lost ; and by continued action of
the cause the underlying parts may become inflamed in
the corium, and the mass is cast off with an undermining
of serous and sometimes hemorrhagic exudation. In those
callous accumulations which form on the flexures of the
joints of the fingers painful cracks often result (Plate 40).

CLAVUS.

Clavus, or corn, is a horny accumulation with a cone-
shaped core or hard center, which is pressed into the cutis,
the apex downward. The formation originally consists
of concentric layers of cells heaped one upon the other,
lies in a sweat-gland duct, and presses upon the cutis,
and may thus cause disappearance of the underlying
papillæ.

CORNU CUTANEUM.

Cornu cutaneum, or cutaneous horn, takes its origin
from the surface of the skin, from apparently fibrous
tissue, and is observed on the scalp, on the brow, and on
the prepuce ; more frequently in the female sex and in
advanced years. The horns are for the most part spiral
and bent, wider at the base, and of a dirty-brown color.
Treatment consists in operative removal of the growth
together with the underlying base.

VERRUCA.

Verruca, or *warts*, are flat, variously elevated, projecting growths of the skin ; they are not sensitive and are seldom smooth, but mostly have a cleft, rugous, dark-gray
surface. They consist of considerably enlarged papillæ
and an increased and hardened epidermis.

The favorite sites are the hands and face, less frequently
the hairy scalp ; it is not uncommon for several to be in

close proximity. **It** is rarely possible to assign a cause for their appearance ; in some cases a persistent irritation of the skin seems a possible **factor.** They sooner or later disappear spontaneously ; or new ones continue to appear, singly or more numerously. Sometimes they appear at the periphery of a group, the central older growths undergoing involution, and in this manner forming irregularly circular areas.

Ordinarily, warts are merely a disfigurement and occasion no discomfort ; but they may become torn and sometimes fissured, and in this way give **rise to** various infection-possibilities.

Treatment.—Warts are removed with the sharp curet or curved scissors, and subsequent cauterization of the base with nitric acid, chromic acid, liquor ferri sesquichloridi, or glacial acetic acid. The growths may also be removed by the thermocautery **or** by electrolysis.

ICHTHYOSIS (FISH-SKIN DISEASE).

Ichthyosis **is a** disease chiefly of the epidermis, dependent upon hereditary disposition. It develops early in life, mostly in the second year. According to type or degree of the disease, several varieties are encountered.

Ichthyosis simplex is observed chiefly on the extensor surfaces of the extremities ; but may also appear upon the trunk. The surface of the skin feels rough, and the small papular—follicular—elevations are covered with firmly adherent scales, upon the removal of which the surface-hairs are observed. This mild type causes the patient very little annoyance.

A more marked type **of the** disease is the so-called *ichthyosis serpentina*, which **is** characterized by dirty-brown, horny scales and scaly plates on the surface of the trunk and extremities. Over the elbows and knees **the** condition is **often** distinctly papillomatous or warty in appearance. The skin of the face is also dry, scaly, and grayish (Plate 39).

The most pronounced grade of ichthyosis is the so-called *ichthyosis hystrix*, in which the affected epidermis consists of polyhedral plates and accumulations, papules, or spines, apparently made up of lamellar and fibrous tissue. There are also numerous markedly enlarged papillæ. The under surface of these spines is uneven, projecting from which are seen hardened papillæ. Not only are these various formations of a dark color, but the skin as a whole also assumes a dirty-gray or brownish hue, so that the patient presents a remarkable appearance (hystricismus). A family of such extreme cases (the Lamberts, father and two sons) was exhibited and described in the last century as " porcupine men."

Unfortunately, ichthyosis is a disease which remains incurable, and in the more severe cases at least, owing to its recognized hereditary tendency, is, with properly-minded people, a hindrance to marriage. The mildest types practically disappear during the heated season, and the more severe cases are also favorably influenced by a warm temperature.

Treatment.—The removal of the scales and horny formations is attained by rubbings and washings with sapo viridis, Wilkinson's ointment, β-naphthol salve, and salicylic-acid-resorcin-tar salves, in combination with baths and prolonged wet packs. In average cases the skin is made smooth and flexible by these measures, and it can be kept in this favorable condition by applications of fat, glycerin-baths, starch-baths, and sweat-baths daily or occasionally, according to the type of disease and the season of the year.

The horny papillomatous outgrowths in ichthyosis hystrix are to be removed by caustics or by operation.

Internal treatment has, up to the present time, proved of no value.

ACANTHOSIS NIGRICANS (KERATOSIS NIGRICANS (KA-POSI) ; DYSTROPHIE PAPILLAIRE ET PIGMENTAIRE OF THE FRENCH).

This rare disease is characterized by two peculiarities—the pigmentation and the papillomatous growths in the skin. Generally the first symptom is the intense dark pigmentation ; it is usually only later that the papillomatous growths are added. The sites of predilection are the neck, the axillæ, the breast, the navel, anal and genitocrural regions, and the poplitea. In isolated cases the mucous membrane of the mouth and the tongue also share in the process. This condition of the skin causes no special trouble beyond the fact of its presence and the disfigurement caused ; but as the disease is usually on covered parts, this latter is of comparative insignificance.

It is worthy of note that in the majority of the cases so far reported carcinomatous disease of the stomach or of the uterus was present, so that the skin-conditions were overshadowed by the symptoms produced by this latter disease. In a case reported by Spietschka there was a deciduoma malignum, after operation for which the skin-affection disappeared.

Histologically, one finds pigmentation, papillary growth, and thickening of the stratum corneum. The pigment is chiefly seated in the basal cylinder-cells, in stratum papillare and subpapillare, and in the lymph-channels of the glands. Changes in the cutis are of an unimportant nature.

The treatment is to be based upon ordinary hygienic rules, modified by circumstances ; the character of the concomitant basic disease indicates that the eventual termination is unfavorable.

PSOROSPERMOSIS FOLLICULARIS VEGETANS (DARIER).

Darier has described an independent disease in which there are growth and hardening of the epidermis, an affection in which apparently the cutis has no share. There appear small horny formations due to hyperplasia in the stratum corneum, which are pointed toward their lower part and project from the epidermis.

These small papules are not only found in the sebaceous gland outlets, but also can be found everywhere in the epidermis. The stratum Malpighii underlying the formations is here and there thinned. Neck, brow, inguinal region, axillæ, and backs of the hands are attacked. The psorosperms which Darier found are not now believed to be in reality these bodies, but arise, according to the latest investigations, through concentric cornification of the epidermis-cells. They are met with in two forms—as rounded little bodies the size of an epidermis-cell, with a nucleus, most abundant in the granular layer ; and as an irregular formation, without nucleus, in the upper epidermis-layers. The acceptance of a parasitic cause for this peculiar dermatosis is, therefore, still an open question.

The disease described by Paget—" *Paget's disease* "—and likewise that reported by White as keratosis follicularis, are looked upon as identical with Darier's disease. Kaposi has remarked that these keratoses remind him of lichen ruber acuminatus.

ANOMALIES OF THE HAIR.

Alopecia.—*Congenital alopecia* is observed mostly as an insufficient hair-growth with lanugo-hairs, which may sometimes be replaced by normal or increased hair-production.

Alopecia senilis is the alopecia coming almost invariably with advancing years, which begins from the brow, extending toward the occiput, the hair still remaining on the sides.

Acquired hair-loss—*alopecia præmatura*—appears between the twentieth and thirtieth years, as a result usually of hereditary predisposition. Frequently this form of alopecia is met with in several members of the same family. As also in the senile form, the hair-loss on the involved region in these cases, with the exception of insignificant lanugo, is complete and permanent. The skin is smooth and shiny, and the follicles are atrophic.

The loss of hair during or following acute disease, as, for example, typhus and typhoid fevers, puerperal fever, syphilis, and inflammatory and parasitic diseases, is, as a rule, temporary.

Alopecia Totalis Præmatura Neurotica (Plate 41, *a*).— Sometimes the hair falls out in young individuals inside of a few days or weeks, without any recognizable disease of the hair. At times it is noticed to be dry, and with a tendency to split or break. The most conspicuous loss is of the hair on the scalp ; but the eyebrows and eyelashes also fall out, and frequently the pubic and axillary hair, and, in fact, the surface-hairs of the entire integument.

The hair that sometimes grows after such loss is thin and atrophic, and soon falls out. The skin shows no changes worth mentioning. It is to be noted that frequently the nails share in the disease, and are milky and fragile. Almost always nervous symptoms are associated, such as nervous disturbances, migraine, and psychoses. The hair-fall in these cases is considered to be a trophoneurosis.

Alopecia Areata (Plate 41, *b*).—Without apparent skin-changes bald spots appear on the scalp, which peripherally enlarge ; frequently only one or several at a time. The hairs seem of normal appearance. Those which are at the immediate periphery of the patches, as a rule, may be easily pulled out. The skin is pale, but without change in the sensibility. Frequently contiguous bald spots become confluent, and there then arise larger hairless areas ; seldom, however, complete baldness of the entire scalp. After some months lanugo begin to appear, which later are replaced by normal hair. Such regrowth usually takes place in from one to two years, sometimes in a shorter period. In spite of the fact that distinguished dermatologists have given the etiology of this disease considerable study, as yet there is no uniform view as to its cause. Some consider the disease a trophoneurosis, others believe it to be parasitic. Since for both these views there is much evidence, the opinion of Lassar seems the correct one—that there are etiologically several processes with the same clinical picture.

Treatment of alopecia naturally should be based upon what seems to be the possible etiologic factors in the case under consideration. It must, however, be remarked that this theoretic division in the treatment is not closely followed, but that substantially all therapeutic efforts have in view a local irritation, and the various local remedies employed to produce this are of stimulating and antiparasitic character. For the neurotic type internal tonics are especially recommended, such as iron, arsenical preparations, pilocarpin, local massage, application of the faradic and galvanic currents. Of the local remedies, may be named the application of salt solutions, acetic acid, tincture of cantharides, tar tincture, oil of mace, chrysarobin, resorcin, etc. If seborrhea is associated, it must be treated according to approved methods, as this condition has an important etiologic bearing in such cases.

Trichorrhexis nodosa occurs more commonly on the bearded region, as nodular excrescences on the hairs. On the hair-shaft may be seen one or several such swellings. The hairs break easily at these points, and there remains a brush-like extremity (Plate 64). Hodara[1] states that he has found a microörganism in this disease and has cultivated it; he was able to produce on sound hairs the same disease. Spiegler has also had a like experience.

Treatment.—Treatment is usually fruitless. Besnier advises depilation of the diseased hairs and the application of tincture of cantharides. Eichhoff advises keeping the hair closely cut and the rubbing in of

R̸ Vanillini,	0.20 (gr. iij);
Adipis,	10 (ʒiiss).—M.
Ft. unguent.	

[1] *Archiv für Dermatologie und Syphilis*, Bd. 41, I.

Tar-sulphur salves, **aqua** ammoniæ, etc. have also been recommended.

Sometimes apparently normal **hairs** are seen with the ends split (**trichoptilosis**). For this condition **dryness of the hair** has been considered responsible.

The term **hypertrichosis** (excessive hair-growth) signifies not only that hairs may appear on unusual situations, but also that hairs in normal regions may be unusually long and thick. Thickness and length of the individual hairs are often associated with luxuriant growth.

The hair in **albinismus** is absolutely without pigment, yellowish-white, soft, thin, and of silky appearance. As acquired, we frequently see whitening of the hair on colorless skin-areas; frequently, however, also without the loss of pigment in the skin (Plate 41).

Graying of the hair depends (Ehrmann) upon the want of pigment-bearing cells in **rete** and in the hair-bulb. **One** finds hairs which may **have a** dark end-portion and an already whitish, grayish shaft. *Sudden graying of the hair*, which is stated to occur from fright, and which is said to be due to the formation of gas in the hair-shaft, needs scientific confirmation.

ANOMALIES OF THE NAILS.

Irregular formation and shapes of the nails arise from excessive growth, by thickening and malformation in consequence of hypertrophy of the nail-bed. The nails become claw- or talon-like, and twisted **like a** horn (onychogryphosis). The latter arises from **the fact** that the nail-body is lifted up from the nail-bed by the accumulation of hardened masses beneath (Plate 40); **or** such growth and accumulation may take place at the forward part of the nail-bed only. The borders are hypertrophied and the lamellar masses show a structure similar to that of cutaneous horns.

Absence of the nails is observed after paronychia, **in**

9

atrophic conditions of the end-phalanges, and possibly in disease of the neighboring epidermis (psoriasis, etc.).

Digestive disturbance and chronic intoxications, acute infections with recurrences, as erysipelas, local irritations (ill-fitting shoes), and inflammation of the surrounding parts may act as causes of hypertrophy of the nails. Diseases of the skin, as chronic eczema, psoriasis, lichen ruber, elephantiasis, syphilis, ichthyosis—in short, all those which are attended with cell-infiltration of the papillary layer and a hyperplasia of the epidermis may act as factors in the production of onychogryphosis; and also the parasitic diseases of the nails, as favus and tinea trichophytina.

PIGMENT-ANOMALIES OF THE SKIN.

Paleness or **whitening of the skin** occurs in anemic states, in consequence of lack of blood after hemorrhages, after depressing diseases, and in chlorosis and dropsy.

Congenital want of pigment (*albinismus universalis*) occurs as a hereditary anomaly. The otherwise normal skin of such individuals (*albinos*) is completely without pigment, white, pinkish, or reddish in color. In consequence of the blood-vessels shining through the iris the eye appears red. Albinos, as a result of the lack of pigment, are sensitive to light and have nystagmus. The hairs are fine, silky, shining, and completely white.

Also after certain diseases of the skin the pigment normally present disappears completely.

Albinismus partialis occurs as congenital loss of pigment in circumscribed regions of the skin. Its distribution frequently corresponds to the domain of a nerve, and, unlike acquired pigment-atrophy, the areas are surrounded by normally pigmented skin. The hairs in such pigmentless regions may also remain white; this, however, is not invariably the case.

The acquired form of pigment-loss (*vitiligo, leucoderma,*

leucoderma acquisitum) (Plate 41) begins at first as small white spots, which spread slowly and irregularly; the bordering skin is overpigmented. There is no **text**ural change in the skin of such areas beyond the loss of coloring-matter, the integument being otherwise anatomically normal; moreover, there are no functional disturbances. The disease may in **the course** of years involve almost the entire surface, **a** few **dark** stripes or areas being left. The hairs become white with the skin. Innervation-disturbances have been looked upon as responsible for this variety. It is only occasionally that an external factor may be productive of these spots, as, for example, pressure of bandages or constriction of scars. **It is** known that after certain acute diseases, as, **for example**, typhus, scarlatina, etc., vitiligo has been observed **to occur.** In most cases, however, the affection is seen between the tenth and thirtieth years in individuals apparently otherwise **in normal health.** The investigations of Ehrmann, Jarisch, Riehl, **and** others have shown how **the** displacement of pigment takes place by means of cells, without, however, throwing light **on the** actual causes of **the** process.

Increase in pigmentation may occur **as a** congenital condition; it is, however, more frequently an acquired affection.

Brownish, brown, and black discoloration **of the** skin, in variously-sized areas, is observed **as a** congenital affection—pigmentary moles (*nævus pigmentosus*) (Plates 35 and 36). Small moles may also be an acquired blemish.

On several regions of the body are observed circumscribed pigment-spots, such as *freckles* (*lentigo, ephilides*). They are, as well known, millet-seed-sized to pea-sized, or possibly larger, yellowish-brown or brownish in color, which are met with in summer on the face and on the hands, but occasionally also on parts covered with clothing, disappearing partly or completely in winter-time.

Of other varieties of pigmentation, there remains to

be mentioned that which sometimes occurs in association with diseases of the female sexual organs—the so-called *chloasma uterinum*, a yellowish and grayish or brownish discoloration on the face, on the areola of the nipple, and in the linea alba. Discoloration of the buttocks, trunk, and extremities was observed in cystic degeneration of both ovaries (Neusser); after double ovariectomy the discoloration rapidly disappeared.

Treatment.—Of prophylactic importance in lentigo is avoidance of the sun's rays during the summer season. In persistent freckles and also in chloasma and other discolorations the application of corrosive-sublimate solution, alcoholic or aqueous, $\frac{1}{2}$ to 1 per cent. strength, is to be recommended. Covering the affected areas with compresses wet with the solution is useful; its action should be carefully watched if the stronger solutions are employed. Also the application of the following:

    ℞ Bismuthi subnitratis,
        Hydrarg. præcip. alb.,   *āā* 5 (gr. lxxv);
        Adipis,   50 (ℨiss).—M.
    Ft. unguentum.

Or salves of β-naphthol or resorcin, already referred to.

In addition to these several methods for the removal of freckles and chloasma, may also be mentioned the application of:

    ℞ Adipis lanæ,   5 (gr. lxxv);
        Vaselini,   10 (ℨiiss);
        Hydrog. peroxid.,   20 (fℨv);
        Hydrarg. chlorid. corros., 0.05 (gr. $\frac{3}{4}$);
        Bismuthi oxychlorid.,   0.5 (gr. viiss).—M.
    Ft. unguentum (Unna).

Leloir advises washing the parts with sapo viridis or alcohol, and then painting on a 15 per cent. solution of chrysarobin in chloroform; the spots, after this dries on, are painted with a solution of gutta-percha. Hardy recommends the application of :

R̸ Hydrarg. chlorid. corros.,   1 (gr. xv);
Zinci sulphat.,
Plumb. acetat.,   āā 2 (gr. xxx);
Aquæ destillat.,   250 (f̃ ℥viij).—M.
    Sig.—For external use.

In many of the cases of acquired pigmentation large areas, or indeed the entire surface, may be more or less pigmented. The pigment often arises from hyperemia, and this usually from some direct irritation of the skin. On the other hand, in some cases pigmentation results as a consequence of diseases of one or more organs ; in such the pigment may also be deposited within the viscera as well as in the skin. The several diseases or conditions which lead to pigmentation are :

Melasma is a discoloration of large areas, frequently on the lower extremities, widespread, brownish in color, following chronic inflammations and congestions in cachectic and emaciated individuals of impaired nutrition. A like condition, consisting of a general darkening of the skin, is observed in consequence of neglect in individuals with flabby panniculus (chloasma cachecticorum), as, for example, in phthisies, in whom the skin appears greasy, smeary, and discolored.

Vagabonds' disease (Vogt) is a melanosis observed in tramps, which arises from neglect of the skin, lice, and heat.

A dirty-gray discoloration of the skin is observed in malarial cachexias.

The discoloration in *pellagra*, a disease which has been described in connection with the erythemas, may also be mentioned here.

Further, *melanoicterus* of the skin is observed in cirrhosis of the liver and in chronic forms of icterus gravis. It may assume on some regions of the body a peculiar bronze color.

In **diabetes** mellitus melanodermic conditions (diabète bronze **of the** French) **are** likewise observed. Also in **those cases with** which **are associated** polydipsia, polyphagia, pol**yuria, and** glycosuria, **the** pigment-accumulation being present **in** the various organs and lymphglands as well **as** in **the skin.** This pigment contains iron, and **is to be looked upon** as a derivative of **hemo-**globin.

In this group belong also **the** melanodermata observed **in** affections of the pancreas, **with** or without associated glycosuria.

Morbus Addisonii.—This disease, described by Addison, associated with disease of the suprarenal capsules, occurs **as a** bronzing of **the** skin, expressing itself also **in disturbances** of the digestive tract **and nervous** system, and almost always ending fatally. According to Lewin, disease **of** the suprarenal capsules **is** observed in 88 per **cent. of** the typical **cases.** The **discoloration** appears some **time after** the patient has been complaining of feelings of weakness, depression, and **the** sensation **of** pressure in the stomach, often pain in the entire abdomen, increased thirst, nausea, **etc.**

At first the color is a dirty-yellow, yellowish-brown, **or** smoke-gray, **and by** gradual darkening it becomes **that of** bronze, and may **even become** black. The uncovered parts and parts **which** are subjected to pressure of the clothing are most conspicuously involved; sometimes also the mucous membranes of the lips and mouth.

The discoloration is either spread over larger areas, in which clear-white spots are irregularly scattered, or it may appear **in** the form of single irregular patches. The hairy parts may also **be** discolored; the hair itself does not, however, usually share in the process. The skin of the face is, **as a rule, the darkest;** the **nails** and the nail-beds are seldom pigmented.

The skin is smooth and elastic **to the** touch, and inclined **to sweat,** but shows no other **changes** worthy of mention.

The pigmentation arises, according to Neusser,[1] through the medium of the general and local sympathetic nerves; the impairment or abolition of the function of the suprarenal capsules being the underlying factor.

To the general weakness are added depression of spirits, ill temper, and impairment of the intelligence. Emaciation, cachexia, weakness of heart-action, palpitation, and dyspnea are symptoms of the early stages. Death results, with gradual and increasing prostration, in consequence of heart-weakness; sometimes the end comes with high fever, diarrhea, persistent vomiting, delirium, and finally collapse and coma. The duration of the disease varies, the extremes being months and years. Often the end comes suddenly, without the patient having gone through the several stages or symptoms mentioned.

In the preceding remarks processes have been described in which the pigment arises from organic constituents within the patient himself, in consequence of some pathologic process. There are, however, other pigment-deposits observed in the skin, composed of certain mineral substances which have been introduced into the system or skin from without. We will refer to the most important representatives of this group — the pigmentations arising from the use of silver nitrate and arsenic.

The discoloration of the skin from arsenic—*arsenical melanosis, arsenicismus*—occurs after its continued administration or from the fact that the patient's occupation brings him in contact with it (Plate 37).

Arsenic is introduced by the mouth as medicine or is taken unconsciously in drinking-water; or, as already stated, the patient is engaged in some occupation in which arsenic is used. It may also be introduced through the lungs and skin from arsenic-containing carpets, wall-paper, etc.

[1] Neusser, article on "Morbus Addisonii" in Nothnagel's *Pathology.*

The discoloration **appears upon the skin, the mucous** membranes remaining **free.**

The pathogenesis of the pigment-formation is not yet understood. It has been assumed that the poison through its affinity for certain substances breaks up the blood-corpuscles, the blood-coloring-matter producing the skin-pigmentation. The pigment is found in the lowest basal cells of the **rete** and in the **cutis.** The **fact however, as** clinical observation teaches, that the pigment **is** deposited, or often more markedly at least, at the sites of former diseased areas of the skin (eczema, psoriasis, etc.), is not readily explained.

The quantity **of** arsenic **which** may give rise **to** pigmentation of the skin differs materially in different individuals. In the case depicted it appeared after **the** administration of 0.26 (gr. iv) of arsenious acid ; in other **cases only 0.216 (gr.** iij $\frac{1}{4}$) **of arsenious** acid had been taken **; and in one case in** Schrötter's clinic 0.125 (gr. j $\frac{7}{8}$) was sufficient.

As regards the **time required, it has appeared in some** cases after six **months' administration of Fowler's** solution, **in** others not before three years, after **doses of five** to ten **drops** three times daily.

The pigmentation appears gradually, and especially **on those** regions which are normally hyperpigmented. **In most cases** the skin **is noted to** have a **bronze** tint; **not** infrequently, however, a graphite color.

As soon as the arsenic has been discontinued the skin begins to resume its normal hue, especially if other damaging effects upon it by the drug (to be referred to) have not been observed. The more intense the pigmentation and the older the patient the more slowly does it disappear.

As **a** further effect of the administration of **arsenic** on **the skin we have** *arsenical hyperkeratosis* (Wilson). In addition **to the** uniform hyperkeratosis on certain parts, **as** the hands **and** feet, corn-like horny formations appear with central depressions, which correspond to **the** hardened outlets of the sweat-glands. Arsenical hyperkeratosis is said to lead sometimes to the formation of epithelial growths.

Argyria.—By the deposition of reduced silver in the skin from the ingestion of silver nitrate the integument becomes discolored. The silver is found outside of the cells in the finest subdivision. The face is the part most frequently and markedly pigmented ; also the conjunctivæ bulbi become gray, and likewise the nail-bed. In the beginning the skin is pale gray ; after continued administration of the drug it becomes dark blue or cyanotic in color.

As in cases of arsenical pigmentation, the examination of the urine is an important diagnostic help in this disorder ; the presence of silver can be readily demonstrated.

NEW GROWTHS (NEOPLASMATA).

Congenital and acquired connective-tissue new growths are the most numerous of the benign tumors of the skin. They occur as small circumscribed cutaneous excrescences, or as more or less extensive thickenings of the skin.

Verrucous growths belong to the former class. They are hypertrophies of the corium covered with pigmented epidermis. They are either smooth or notched, are often covered with hairs and dilated sebaceous follicles, and are either sessile, provided with a broad base, or pedunculated (Plate 35).

Nevus.—The so-called *nævus mollusciformis* or *nævus lipomatodes* is another example, differing in being rarely sessile and occurring as pendulous, pedunculated small tumors on the skin of the neck and eyelids.

Nævus spilus occurs in the form of elevated lesions, the size of a lentil or bean, but may occasionally be distributed over large cutaneous areas in the form of hypertrophy of the skin and papillæ, accompanied by black or dark-brown pigmentation, which often extends even into the cutis. Nævus spilus is very frequently covered with **stiff** hairs.

Treatment.—In the smaller lesions the application of caustics, such as trichloracetic acid, lactic acid, and nitric acid, or sublimated collodium (5–10 per cent.), is recommended for the removal of flat nevi ; **if** not successful, electrolysis or excision is to be advised.

When the growths are of larger size **and** removal by surgical means is contraindicated **or** objected to, recourse may be had to electrolysis (Voltolini, Hardaway, Fox) **or the** galvanocautery.

Cicatrix or Scar.—Losses of substance of the skin extending into the corium, or **at** least into the papillary layer, are replaced **by** cicatricial formation. Scars are likely to occur after **burns,** suppuration, **and caustic** applications, and after diseases leading to **purulent** destruction **of tissue, as** lupus, scrofulosis, **and various** dermatitides, or extensive hemorrhages **and** gangrene of skin. After the necrotic mass has come away the granulation-tissue to replace the defect begins to form. The proliferation of the granulation-tissue commences in the deeper parts, and is gradually **converted into connective** tissue, which becomes covered with an imperfect epidermis. The numerous islands of epithelium which can be seen in extensive wounds after burns probably originate from the epidermis of the sebaceous and sweat-glands. Fresh cicatrices are rich in cells and contain numerous blood-vessels ; the older ones, however, contract, the blood-vessels become occluded, and fibrous connective tissue forms. Cicatricial formation is frequently of great significance, according to its **extent** and location, as it frequently leads to contraction and fixation **of** the articulations. When involving the face, the orifice of the mouth is distorted or contracted, ectropion of the eyelids results, and finally, owing to constriction, circulatory disturbances frequently supervene, which lead—especially on the extremities—to secondary edematous stases and overgrowth of the tissue, elephantiasis.

Keloid is a flat, elevated, white or bluish-red, firm, tumefied, **cicatricial** hypertrophy, **which** frequently sends out claw-like processes. It is **covered** with a thin, shining epidermis, and consists of accumulated embryonic connective-tissue elements embedded in dense fibrous **tissue.**

Elephantiasis Arabum; Pachydermia.—This disease represents a hyperplasia of the corium and hypertrophy of the papillæ. The enormous size of the affected part sometimes reached is due to marked hypertrophy of the subcutaneous cellular tissue; this condition occurs most frequently on the lower limbs. Higher **degrees,** with irregular hypertrophy and sclerosis of the subcutaneous connective tissue, and various, even verrucous, vegetations of the papillary layer, are met with; occasionally, thickened tuberosities and firm linear infiltration are at first noted beneath the skin. The integument and subcutaneous tissues are permeated with serum, and in very advanced stages the muscles down to the periosteum and subcutaneous tissue are degenerated and indurated (*elephantiasis Arabum*). Eczematous, erysipelatous outbreaks, with inflammation of the connective tissue associated with phlebitis and lymphangitis, is the direct cause of these deformities. Hardening and obliteration of the veins and lymphatic vessels lead to these consecutive phenomena, which may occur not only on the lower, but also on the upper extremities, and on the scrotum and labia. Necrosis of the epidermis overlying these sclerotic masses of connective tissue often occurs, and there result large, sinuous ulcers with perpendicular, callous edges, surrounded by cicatricial tissue, papillomatous vegetations, and eczematous skin.

Scleroderma.—This chronic disease is characterized by board-like consistence and rigidity of the skin. It usually occurs in circumscribed patches on the upper half of the body or diffused over larger areas. It is met with on the face, neck, and upper parts of the chest and back, on the upper extremities, more rarely on the abdomen and lower extremities. The affected skin is firm and hard, and cannot be pinched up. Extension occurs either in irregular patches or in the form of streaks or bands, or diffused over larger areas. The surface is either shiny or of a dull brown-red color; whitish areas alternating with irregular dark-brown pigmented spots. The hands grow livid and of a cyanotic hue. Owing to the skin being bound down firmly and the underlying muscles and joints being tightly encircled, their movements are interfered with. When the skin of the face is involved it has a rigid expression, the mobility of the lips and eyelids is impaired, the nose is contracted, and the whole face—to quote a classical expression of Kaposi's—appears as if petrified and hewn in marble. The articulations of affected extremities can only be slightly moved or not at all; the fingers are semiflexed and rigid. Any attempt to extend the joints and pressure on the skin give rise to pain. Tactile sensation and also the function of the cutaneous secretion are not materially altered; the temperature is somewhat lowered. The disease frequently extends irregularly; it has, however, also been noticed to follow the distribution of the peripheral nerves. Unilateral localization, as in eruptions of zoster, along the various nerves has been described.

The commencement of the disease is often so slight that the patient's attention is only attracted to it by a sensation of tension. Occasionally the process is preceded by muscular and articular pains, or by intense erythema accompanied by edema, which may exist for weeks before the skin becomes sclerosed. In these stages the skin may sooner or later return to the normal condition and the sclerotic foci undergo resolution. The disease at times recurs, the foci increase in extent, and the condition lapses into the so-called atrophic stage of scleroderma, which is not susceptible of improvement. The skin becomes thin and resembles parchment, the follicles become obliterated, and the glands atrophy. The discoloration, the shiny appearance, and the hidebound condition remain unchanged. Owing to pressure, the subcutaneous fat shrinks, and even the muscles atrophy (atrophy of inactivity). These atrophic conditions at times result in disease of the joints and of the periosteum and bones.

The irritation of the skin frequently leads to ulcerations on projecting parts, and even to gangrene.

After the disease has continued for years emaciation and aggravated marasmus ensue. The fatal issue is usually due to intercurrent diseases.

The diagnosis of scleroderma offers no difficulty, for the characteristic changes and the hard, smooth, cool-feeling skin are significant. It might be mistaken for pigmentation of the skin occurring in Addison's disease—in this affection, however, the skin is not sclerotic; or for xeroderma pigmentosum, but in the latter the appearance of carcinoma is characteristic.

The etiology is as yet unexplained. The disease is more common in women. Its origin is frequently attributed to disturbances in the domain of the peripheral nerves; this view has some support, as certain forms, already mentioned, follow the distribution of nerves. Vascular changes, especially compression of the vessels in places, have been described, which circumstance has led to the conclusion that scleroderma is possibly to be regarded as the result of an inflammation.

Treatment.—Nourishing diet, tonics, plain or medicated baths, and the internal administration of potassium iodid and sodium salicylate have been advised. Locally massaging of the diseased parts with an indifferent fat or salicylic-acid ointment is probably the most efficient procedure. Some authors laud the action of the constant electric current.

Scleroderma Neonatorum.—Induration of cellular tissue in new-born infants usually appears during the first months of life; it begins with edema and sclerosis of the feet and lower extremities, and gradually spreads in a few days over the rest of the body. The temperature declines steadily, and death usually results in two to ten days. The affection is frequently associated with cardiac lesions and diseases of the respiratory and digestive tracts in weak or debilitated children.

Myxedema.—This disfiguring affection occurs principally in females, and consists of increase in volume of the affected cutaneous parts, which appear swollen, thickened, and hardened. It is met with on the face, on the trunk and extremities, and also on the tongue and velum palati. The hands and fingers also appear more or less deformed, owing to thickening.

The mental and physical faculties of such individuals are also impaired; mental hebetude ensues, the senses of taste and smell are lost, and they are incapable of physical or mental labor. They frequently perish of cardiac and renal disease.

The affection is due to proliferation and deposit of mucin in the skin, in the muscles, and also in the internal organs.

Œdema cutis, or anasarca, due to circulatory disturbances, is allied to this process. It represents a secondary phenomenon, and not an individual skin-disease.

PARTIAL ATROPHY AND THINNING OF THE SKIN.

This condition occurs most usually during middle life in hydrops, anasarca, pregnancy, and rapid accumulation of fat, owing to tension and stretching of the skin; the deeper tissue-layers are spread apart and the skin becomes very thin (*striæ gravidarum*). The streaks at first are bluish-red; later they turn white and shiny, and resemble cicatrices (*striæ atrophicæ*).

Pressure from a bandage or from an internal tumor causes the skin to become hyperemic for a time; the macerated epidermis desquamates freely, and finally the skin may atrophy, and after persistent pressure cutaneous necrosis or ulceration and disintegration may result.

GENERAL ATROPHY OF THE SKIN.

This occurs during advanced age as a degenerative involution of the skin and its appendages. Diffuse progressive atrophy of the skin is furthermore induced by many as yet unexplained pathologic processes. The atrophic skin is exceedingly thin and wrinkled, resembling cigarette-paper. It is inelastic, and when pinched into folds returns slowly to its original shape. The veins are dilated, and can be seen as bluish lines shining through the thin, translucent epidermis. The secretion of the sweat-glands continues in but few places, as the genitalia, face, and axillæ. The hairs are lost; only a lanugo-hair here and there is still visible. The process must be regarded clinically as an atrophy. The progressive form of atrophy of the skin has been demonstrated histologically to be preceded by a chronic inflammatory process, which takes place principally in the layers of the cutis. The sequelæ are shrinking and atrophy of the papillary layer and of the sebaceous and sudoriparous glands and hair, and increase of connective tissue in the deeper parts of the cutis.

XERODERMA PIGMENTOSUM (XERODERMA, PARCHMENT-SKIN).

Kaposi was the first to describe this malady and to call attention to its malignant character. The disease develops in consequence of congenital predisposition in early childhood, rarely later, and the main characteristics are yellowish-brown pigmented spots, resembling freckles; interspersed among these are small telangiectases and slightly-depressed whitish areas, frequently resembling the scars of small-pox. The skin appears atrophic and dried up, resembling parchment, and is tense and can only be pinched into folds with difficulty. The telangiectases are either punctiform or linear. The dilated vessels and pigmentation, and the whitish cicatrices, impart a spotted color to the skin.

The integument of the face, neck, dorsal surfaces of the hands, the forearms, shoulders, and trunk, more rarely the lower extremities and dorsal surfaces of the feet, are involved.

During the further progress of the disease the small vessels are obliterated, and white, shiny, atrophic little depressions and later diffuse shrinking of the skin are to be noticed. As the epidermis also atrophies and exfoliates in the form of lamellæ and becomes fissured, much disfigurement ensues, such as superficial rhagades and ulcers, narrowing of the nasal and oral cavities, and eversion of the lower eyelids.

The rapid spread and continuous atrophic transformation of tissue distinguish this disease from ordinary freckles and pigmented nevi. The vascular changes, consisting of new growth and obliteration, overgrowth of the endothelium, the pigmentary deposit, and projection of the rete downward, and the atrophic processes, are the precursory stages, which stamp this as a peculiar disease subsequently developing into carcinomata and sarcomata.

These malignant new growths may occur in the course of a few months in various places, as the face and external parts of the ears. When this takes place the doom of such patients is sealed, as a fatal termination is inevitable. The epithelial carcinomata appear as warty formations; they increase in size, disintegrate, and soon lead to cachexia and death. The early appearance of xeroderma in childhood and in several members of one family seems to point to heredity.

The treatment of xeroderma has not as yet given positive results. It is usually restricted to symptomatic measures, as may seem necessary, or to operative procedures; without, however, being able to promise much to patients.

Kaposi mentions, as a second form of xeroderma, an atrophy of the skin of the extremities, which is said to begin in earliest infancy and is only distinguished from atrophic scleroderma by its early appearance.

LUPUS ERYTHEMATOSUS.

This **chronic**, inflammatory disease appears **at first** principally **as** small, **raised, dark-red** spots, which are usually shiny, and the center **covered** with a thin, **adherent**, small scale. In the primary stage the spots extend peripherally, and this extension leads to the so-called *lupus erythematosus discoides* (Plate 42).

These discs, the size of a dime to that of a dollar **or** the palm of the hand, usually occur at first on the bridge **and tip of** the nose, the alæ, **and** cheeks. The shape **of** the discs varies according **as** the peripheral extension **is** regular **or** irregular. **One** of the **most usual varieties** is the so-called butterfly-form, **which spreads from the** bridge of the nose to the alæ and even to **the cheeks.** The **center** of the fully-developed patches **is** depressed, shiny, and cicatricial, and **is** either red **or** traversed by dilated vessels. The margin is redder, elevated, more succulent, and is often covered with scales **or crusts;** these latter are the result of marked exudation from the dilated vessels, the exudation and epidermis drying upon the surface. The inflammation begins principally in **the** follicles and sweat-glands and spreads in the cutis, and extends downward to the subcutaneous cellular tissue as well as upward to the epidermis. The exudation loosens the epidermis, and the latter exfoliates in the shape of small scales, which are adherent at first. Involution takes place in this manner: The newly-formed connective tissue shrinks, atrophic scarring results, the affected areas are depressed and contract, **and** the cutaneous follicles are obliterated; the sebaceous and sweat-glands disappear.

Another form of lupus erythematosus **is** known **as** *lupus erythematosus disseminatus* (Plate 43). In this variety the efflorescences **are** more numerous and make their appearance about **the** same time. Numerous patches or areas **are** scattered **over** the **entire** face and ears. The spots are dark red, slightly elevated, firm and elastic, and

10

the overlying epidermis is fissured, exfoliates, and is studded with dilated follicular openings. When these efflorescences occur over the fingers and forearms the color is apt to be darker and they are firmer than those on the face. In a few instances the hairy scalp and the mucous membrane of the mouth have been found to be involved. According to our experience, such extensive spread of the disease must be regarded as rare.

The course of lupus erythematosus in both forms is exceedingly chronic, the affected areas remaining long unchanged, sometimes for years.

Although the prognosis, even under such conditions, cannot be regarded as absolutely unfavorable, as lupus erythematosus may either undergo rapid involution or may terminate in cicatricial formation and slight vascular dilatation, experience teaches that many of these cases eventually die of pneumonia and tuberculosis. [As observed in this country, in some cases the patches may retrogress and disappear without leaving a trace, new areas usually appearing from time to time.—Ed.]

According to Kaposi's observation, which is not to be underestimated, the greatest number of patients are females, who suffer not infrequently from chlorosis, dysmenorrhea, catarrh of the apices of the lungs, and incipient tuberculosis; male patients, however, appear to enjoy better health.

We have seen a case of disseminated lupus of rather acute character spread rapidly, accompanied by pronounced disturbances of the general health, and the patient, a female, died six months later of an acute pulmonary affection.

Treatment.—In mild cases washing with soft soap or tinctura saponis viridis will at times be sufficient to cause the efflorescences to disappear. The application of salicylic-acid plaster or gray plaster to the diseased areas is also to be recommended. Schütz recommends painting with :

℞ Liq. potass. arsenit.,   4 (f ℥j);
   Aq. destillatæ,   30 (f ℥viiss);
   Chloroformi,   (gtt. ij).—M.
Sig.—For external use.

In obstinate cases recourse must be had to reducing (de-oxidizing) remedies—resorcin, salicylic acid, pyrogallol, etc.—to bring about results. We have had especially good effects with Lassar's method of producing exfoliation, described under acne (*vide* p. 52). Galvanocautery and thermocautery have also been warmly recommended. Multiple scarification followed by dusting with iodoform gives good results (Veiel).

Growths of the connective tissue, in the narrower sense of the word, have a more projecting character than the diseases just described, partaking of the nature of tumors. They are:

Fibroma Molluscum seu Pendulum (Fibroma). —Fibromata consist of rounded, usually pendulous tumors, rarely flat, and provided with a broad base; they feel doughy, lobulated, soft, or somewhat firm to the touch, and are invested with normal skin. They generally occur on the head and rump, but in some cases hundreds of varying size may be scattered over the entire body. The larger growths prove annoying owing to tension, interference with motion, and occasionally the occurrence of inflammation or even gangrene of the overlying integument. They are regarded as hyperplasias of the connective tissue taking origin in the deeper layers of the corium or nerve-sheaths (neurofibromata), and consist at first of gelatinoid, later on of fibrous connective tissue. The vessels are contained in the pedicle. The skin adheres to the distal end of the growth, and consequently represents a pouch in which the tumor is suspended.

According to the tension and distortion which may exist, various changes in the glands and epithelial investment occasionally result, and owing to consequent inflammation, ulceration, and gangrene the tumor falls off. Spontaneous involution has also been observed.

Hereditary predisposition is regarded as a causative factor, inasmuch as the tumors are frequently observed during early life. Hebra has pointed out that the patients usually are degenerates in body and mind.

Treatment is exclusively surgical.

Lipoma.—Lipomata, or fatty growths, do not appear before advanced life, and form lobulated, soft, elastic tumors. They are usually multiple, and either have a broad base or are provided with a pedicle and are pendulous. The overlying skin is normal in appearance, and is seldom changed by distortion and traction, as is the case with fibromata.

The treatment of lipomata is surgical.

Xanthoma; Xanthelasma; Vitiligoidea.—Xanthomata are sharply-defined, flat, slightly-raised or tuberous small plates, projecting from the skin. The former (**xanthoma planum**) are spots of a yellow or chamois-leather-yellow color; they are of soft consistence and usually occur on the internal or external canthi; the ears, nose, and even the mucous membrane of the mouth may also be the seat of the growths. They appear in women about the climacteric, but also in men of more advanced years, without causing annoyance apart from the disfigurement.

The second variety, **xanthoma tuberosum**, occurs in the form of tumors the size of a pin-head to that of a hazel-nut. They are of firmer consistence and of irregular, lobulated construction. The lesions are red at the base and yellowish at the apex. The tumors occur on the extensor surfaces of the joints, the fingers, elbows, knees, on the nape of the neck, and in the sacral and gluteal regions. They have also been found on the mucous membranes and even in the internal organs (endocardium, wall of the aorta, etc.).

Xanthoma represents anatomically a connective-tissue tumor with interspersed specific xanthoma-cells. The etiology **of xanthomatosis is as yet** not known. Frequently jaundice, disease of the **liver, or** diabetes co-exists. The latter disease **especially** appears to predispose to xanthoma tuberosum. In our case (Plates 44 **and** 44, *a*) striking involution **of** xanthoma-tubercles occurred twice after the disappearance of **diabetes** had been brought about by a bath-course at Carlsbad. The disease relapsed simultaneously with the appearance of sugar in the urine, and at both times the small tubercles involuted when sugar ceased to appear in the urine.

Treatment.—Xanthoma planum is most readily **re**moved by surgical procedures [also by electrolysis—ED.]. In multiple eruptions of xanthoma the general condition **must be** carefully looked into ; patients must be examined **for diseases** of the liver, gout, diabetes, glycosuria, and **nephritis. The** xanthoma-tubercles have been repeatedly **observed to** retrograde under proper general treatment. Broeq recommends the internal use of phosphorated **oil** and oil of turpentine.

Dermatomyomata.—Myomata are rare skin-lesions. They occur around the nipples, on the scrotum and ex-**tensor** surface of the arms ; are firm pea-sized tumors, which **are** movable with the skin. The overlying skin is more pigmented than usual ; otherwise **it** remains **un**changed.

The tumors are not, **as a** rule, painful **on** pressure, although **some** have been described **as** being spontaneously painful. They develop from **the** smooth muscular fibers or from the enveloping **or** immediately adjacent connective tissue. They start from the arrectores pilorum. In a few cases numerous vascular coils and nerves have been found along with the hyperplasia of the cells of the muscles, and such **cases** possibly represent the especially painful little tumors.

ANGIOMATA.

A. **Nævus vasculosus** is most commonly a congenital dilatation of the capillaries and smaller cutaneous blood-vessels, and is usually on a level with the skin. The color of the vascular nevus is mostly dark red or bluish-red, and depends on the predominance of the dilatation of large vessels or small capillaries. Dilatations the size of a split pea are frequently seen scattered irregularly over the trunk; larger nævi vasculosi occur on the face (temporal region), hairy margin, nape of neck (Plate 45, *a*), and even scattered over larger portions of the body-surface (Plate 45). The larger vascular nevi occur unilaterally and may increase in breadth. The anxiety of mothers, therefore, to have the small vascular nevi in newly-born children removed as soon as possible is not without foundation. *Telangiectases* are acquired blood-vessel new formations, usually consisting of enlarged capillaries or a pin-head to pea-sized dilatation, with or without enlarged capillaries extending from it. Vascular dilatations resulting from venous stasis due to interference with the return-circulation have been considered in their proper place. Venous dilatations often form plexuses the size of an egg to that of a fist, which are very troublesome, as they frequently lead to inflammation. They occur on the lower extremities, in the spermatic plexus, and in hemorrhoidal veins, and must be removed by operation (Dittel's elastic ligature, hot-wire loop, or excision).

B. **Lymphangioma.**—The capillary lymphatic vessels of the skin are dilated owing to interference with the flow of lymph either by infiltration and occlusion of the larger vessels or by swelling of the lymphatic glands in whose domain the lymphatic vessels are situated. Should any of these small lymphatic vessels rupture the lymphatic fluid oozes forth continuously. Pressure of a bandage may also lead to decided dilatations of the lymphatic vessels.

More extensive dilatations of the lymphatics are observed as nodular formations in a swollen area of the skin, and exhibit not only varicosities and dilatations, but many new vessels form in the corium. We

have observed such swellings on the scrotum and penis.
Finally, dilatations of the lymphatic vessels, accompanied
by swelling and hypertrophy of the skin, often over a
whole region of the body, occur, especially on the lower
extremities; these are known as *elephantiasis lymphan-
giectodes*, and, with accompanying blood-vessel hypertrophy, closely resemble ordinary elephantiasis.

Lymphangioma Tuberosum Multiplex. — Kaposi and others have described numerous, partly round,
partly elongate, brown-red nodules lying in and movable
with the skin, situated on the trunk and region of the
neck. Not having any personal experience with this
rare skin-disease, we refer the reader to Kaposi's treatise
on skin-diseases.

RHINOSCLEROMA.

The peculiar disease called rhinoscleroma was described by Hebra and Kaposi in 1870. The affection
attacks the nose and spreads very slowly over the skin
and cartilages of this organ and neighboring parts. It
may further involve the posterior part of the soft palate,
the isthmus, larynx, and trachea. The disease spreads
only by contiguity from the starting-point. Rhinoscleroma attacks individuals about the period of puberty.
The patients are usually not robust. Although it cannot
be regarded as a specific hereditary disease, a certain
predisposition is generally thought to exist. One of
the alæ or the septum is attacked by the disease and
the shape of the nose changes gradually without exhibiting decided signs of inflammation. The nose widens and
feels rigid and immobile to the touch. Owing to hypertrophy of the inner walls, stenoses and even complete
occlusion of the nares occur. After months the whole
organ, anteriorly as far as the lips and posteriorly as far as
the choanæ, becomes involved. The external picture
varies, and depends on the presence of tuberosities projecting over the level of the skin or on the presence of uni-

form hypertrophy of the skin and cartilages, resembling plaques. The color may be of various shades of red, but is usually brownish- or bluish-red. Blood-vessels are seen running over the surface, which is smooth or finely wrinkled, and shiny. In the same manner as stenosis of the nasal cavity results, the functions of the lips are also interfered with. We also meet with various distortions and constrictions in the isthmus faucium, which not infrequently remind one of syphilitic sequelæ.

The patient's appearance suffers considerably, and the resulting occlusion of the nose and stenosis of the entrance to the larynx and mouth are a source of great annoyance. The diseased areas are sensitive to pressure. The affection is chronic, extending over years, without necessarily any change in the general health.

Most observers regard the disease as inflammatory, in which the infiltration is partly absorbed and partly converted into connective tissue.

Specific bacilli have always been found in the tissue of rhinoscleroma since Frisch called attention to their presence. Paltauf and Eiselsberg found capsulated bacilli in protoplasmic masses, which correspond to the cells of rhinoscleroma or degenerated nuclei, first described by Mikulicz. The rhinoscleroma-microörganisms appear as 2–3 μ long bacilli, or as ovoid, nearly round, capsulated cocci, occurring usually as diplococci, which can scarcely be distinguished from pneumonia-cocci.

The prognosis is unfavorable; it is impossible to stop the process by any treatment. Surgical procedure is indicated when adhesions and hypertrophy have advanced so far as to interfere with the functions of the parts.

TUBERCULOUS DISEASES OF THE SKIN.

In this section we embrace those pathologic changes in the skin (*vide* Plates 46 to 51) due to the tubercle-bacillus; they show great variety in appearance, course, structure, and pathogenesis. To avoid repetitions, we will follow in this section the classical work of Jadassohn (Lubarsch and Ostertag, 1896); we will, however, first briefly touch on a few general points.

The bacillus may gain entrance to the skin in various ways. External tuberculous material may be either implanted (exogenous inoculation-tuberculosis) or the material originates from an already diseased body—e. g., sputum, saliva, feces, and urine (tuberculosis due to auto-inoculation). Certain external predisposing causes, however, are necessary for the tubercle-bacillus to establish itself, inasmuch as the skin does not appear to be favorably disposed to tuberculous disease. The tubercle-bacillus may find such points of attack where the skin is injured or where cutaneous disease exists—briefly, when wounds of the integument are present. On the other hand, tuberculosis may find its way from a neighboring organ into the skin (tuberculosis due to contiguity)—e. g., from a primarily-diseased testicle to the scrotum, or from bone to the overlying soft parts. Finally, the bacillus may gain access to the skin from a diseased organ by metastasis.

We differentiate clinically five forms of cutaneous tuberculosis :

A. Lupus;

B. Scrofuloderma ;

C. The tuberculous ulcer ;

D. Tuberculosis verrucosa cutis ;

E. Tuberculosis fungosa.

Although it would be gratifying if the clinical symptoms of the above-named forms of cutaneous tuberculosis were always distinctive, the fact must be emphasized that several varieties may exist alongside of one another, and that frequently one develops from another. Thus, for instance, tuberculosis verrucosa may change into lupus ; lupus may develop from a scrofuloderma which has already cicatrized (Riehl). Along with tuberculous ulcers subcutaneous nodules of scrofuloderma, etc. develop. Taken as a whole, however, this, as with other multiform diseases, usually presents in different individuals a distinct type or variety.

All these varieties may terminate spontaneously [exceptional.—Ed.]. The cicatricial formation which in such cases, as in all ulcerative processes, denotes a cure, is, according to the duration and intensity of the disease, at one time slight, at another time more marked, and may lead to shrinking and other consecutive changes. Various authors have called attention to a temporary lull in the course of lupus, and have connected it with possible conditions in the organism itself (pregnancy) or with external influences of temperature and weather.

A. LUPUS.

Lupus is the most frequent form of cutaneous tuberculosis, and occurs principally on the uncovered parts of the body, as the face and hands; and to a less extent on the scalp. It begins in many cases during infancy or early childhood, and is met with more frequently in females.

It has been established that its origin in the greatest number of cases is due to external inoculation; but it may also be conveyed from tuberculosis of the glands and bones, or from diseases of the mucous membrane to the skin. In general, the tubercle-bacilli in lupus are very scanty; usually several are capsulated in the giant cells. The tubercle in the skin consists of round, epithelioid, and giant cells and of a reticulum and vessels. The lupus-nodule represents a conglomeration of such tubercles.

Lupus begins clinically with the appearance of pinhead-to hemp-seed-sized nodules, which are yellowish-gray or brownish-red in color. At first, they are embedded in the skin and project only after they have persisted for some time, and are covered with a smooth, shiny epidermis. The typical nodules at first are also flat and isolated; the various clinical pictures of the disease are due to the changes which take place—to the lesions becoming contiguous and confluent, etc.

The disease soon begins to spread in areas ; the isolated foci are usually sharply defined at the periphery and are surrounded by inflammatory infiltrated cutaneous tissue ; the round-cell accumulation and infiltration follow the vessels. This massed cellular infiltration involves all the cutaneous parts. The elastic fibers, hair-follicles, sebaceous and sudoriparous glands are either destroyed or only their débris remains. The changes in the epidermis are connected with the processes in the cutis ; once we saw rapid death of the epithelium take place ; but usually the inflammation and irritation lead to hypertrophy of the epithelium, especially of the epithelial cones extending toward the cutis. The surface of the lupus-foci is either smooth or covered with scales ; or hyperkeratosis is noted, giving rise to superficial verrucosities.

We usually meet with sclerosis of the inflammatory infiltrate around the lupus-tissue, which leads to absorption accompanied by cicatricial formation. Less frequently, and only on account of special causes (spread of inflammation, secondary infection), do breaking-down and ulceration of the lupus-tissue result. Dry caseation is rare in lupus.

Owing to these anatomic changes, to which we have briefly alluded, and especially to the extension—already mentioned—of the lupus-growths, we differentiate clinically *lupus tuberculosus*, when shiny nodules are either disseminated or irregularly grouped ; or when arranged in rows or closely crowded, and continuing to spread serpiginously (*lupus serpiginosus*) and protruding over the level of the skin ; *lupus tumidus*, when the lupus-growths take on the form of tumor-formations ; *lupus verrucosus*, *lupus papillomatosus*, when the surface appears papillomatous or warty ; and finally *lupus exulcerans* (Plates 46, 47, 47, c).

The lupus-ulcers are usually covered with dark-colored

crusts, the ulcerating surface underneath, on a level with
the skin, appears red and moist, bleeds readily, and resembles granulating wounds.

The mucous membranes **of the nose and oral cavity**
may be the seat of lupus-nodules for a long **time** without
giving the patient especial annoyance.   They are met
with on the gums, palate, tongue, and larynx, as brown-
red, usually ulcerating and readily-bleeding nodules, the
size of a pinhead to that of a split pea.   When they
coalesce and form large plaques the surface is irregular
and covered with gray, proliferating epithelium ; or if
breaking down and disintegrating, form flat or fissured
deep ulcers.

As already mentioned, lupus extends from the mucous
membrane upon the external skin, and *vice versâ*.   Defects
of the palate, due to ulceration and shrinking, and also
depressed contractions of the tongue—the latter are frequently associated with firm nodular swelling in the
neighborhood—are of not unusual occurrence.   We have
often found polypoid vegetations in the nasal cavity along
with ulceration and crust-formation, completely closing
the affected half of the **nose.**   These are distinguished
from translucent mucoid polypi by their granulating **sur-**
face and by their tendency to bleed ; mucoid polypi are
covered with a smooth mucous membrane.   Perforation
of the septum, cicatricial contraction, and distortion are
the sequelæ which frequently follow after the disease has
existed for years.

The exterior of the nose, and especially the alæ, are
frequently the points first attacked by lupus.

The disease spreads gradually from the tip to the root
of the nose.   Papillary elevations at **the** margin of the
ulcers, which are continually disintegrating, become cov-
ered with brown crusts and gradually lead to destruction
of the entire cartilaginous and exceptionally of the osseous
structure.   Lupus also extends to the cheeks and often to
the margin **of the lower jaw** and to the neck ; the submaxillary glands are not infrequently diseased at the

same time and consequently suppurate. Lupus tumidus is frequently met with on the lobes of the ears.

Lupus of the eyelids leads to ectropion and consecutive diseases of the bulbus. It is occasionally primary— although rare—on the conjunctiva of the bulbus and extends to the cornea. On the trunk, especially on the nates, we often meet with the papillary, verrucous forms; and with the serpiginous varieties on the extremities. Owing to cicatricial contraction, the articulations become fixed and the parts are deformed and become useless. Deformities of the hands, especially unsightly hypertrophies, are attributable to disease of the lymphatic vessels (Plates 48, 48, *a*, 48, *b*).

Lupus pursues an exceedingly chronic course. Beginning usually between the ages of ten and twenty years, it extends very slowly, retrogressing on one side, and spreading serpiginously at the periphery; undergoes involution— *i. e.*, cicatrizes—often completely to recur again. Owing to mechanical damage or irritation or intercurrent affections, erysipelas, etc. inflammation, disintegration, and ulceration result, which frequently lead to very great destruction of the face, nose, the soft and hard palates, etc. We would add that in lupus erysipelas is especially prone to recurrence.

The cicatricial constrictions interfere with the circulation, and especially when the disease is upon the extremities lead to chronic edema and elephantiasis of the subjacent or peripheral parts. The spread of the tuberculous process along the lymphatic interstices and vessels, owing to the attending inflammation, leads to elephantiasic transformation, which may involve the soft parts and even the bones.

Syphilitic lesions occurring in lupus-infiltration or in a resulting cicatrix may occasionally complicate the disease. This complication, however, does not justify the use of such a term as "lupus syphiliticus," and it should be dropped from terminology as meaningless.

Syphilitic ulcers may, *vice versâ*, owing to infection

with tuberculous material, be converted into tuberculous ulcerations.

Finally, we will briefly refer to the coexistence of carcinoma and tuberculosis. Carcinoma appears more frequently in a lupus-cicatrix than in fresh lupus-tissue. It starts from the rete or from the glandular organs of the skin, and does not originate, as many authors have thought, from transformation of lupus-tissue into carcinoma.

Treatment of Lupus Vulgaris.—Internal treatment must be directed to improving the general condition; any direct influence on the skin-affection from the remedies recommended is not to be expected. Success can only result from local treatment carefully planned. For the purpose of producing destruction of the diseased foci and areas the following methods are used : Volkmann's spoon, thorough scarification of the affected patches (Balmanno-Squire, Vidal), cauterization with Paquelin's cautery, galvanocautery, thermocautery ; excision followed by transplantation—a procedure which in expert hands leads to good results (Lang).

For the purpose of producing destruction by chemical means the *caustic pastes* are employed, as the Vienna paste (quicklime, 4 parts ; dried caustic potash, 5 parts), zinc-chlorid pencils (obtained by fusing zinc chlorid and potassium nitrate, or zinc chlorid and potassium chlorid, with a cover of tinfoil), Canquoin's paste (zinc chlorid and rye flour, equal parts), Landolfi's paste (zinc chlorid, 3 parts ; bromin chlorid, 5 parts ; chlorid of antimony, 1 part). These pastes act on the healthy skin as well as on the diseased skin.

Cosme's paste (arsen. alb., 1 (gr. xv) ; cinnabar. factitiæ, 3 (gr. xlv) ; ung. emollient., 24 (ʒvj)) acts by election— i. e., it destroys the lupus-nodules, but leaves the neighboring healthy skin intact. Elective action can, of course, be also obtained with the silver-nitrate stick, and with cauterization with carbolic, lactic, and pyrogallic acids.

We have to note very satisfactory results with a 20–25

per cent. ointment of pyrogallic acid. The pain which the application generally produces is not great, and may be diminished by adding orthoform. After several days or longer the formation of the eschar in the lupus-infiltrations is complete ; separation and cicatrization are allowed to terminate under an indifferent ointment (boric-acid ointment).

According to the Unna-Scharf method, sharpened pieces of wood (toothpicks, shoemaker's pegs, etc.) which have been lying for a few days in the following solution are introduced into the lupus-foci :

R, Hydrarg. chlorid. corros.,   1 (gr. xv);
    Acidi salicylici,   10 (ℨiiss);
    Æther. sulph.,   25 (fℨvj gr. xv);
    Ol. olivæ,   ad 100 (fℨiij ℨj).

All the wooden stumps projecting over the level of the skin are then cut off with scissors and the surface thus treated is covered with any kind of gutta-percha plaster ; the best is Unna's gutta-percha plaster of mercury and carbolic acid. After removal of the plaster the surface is seen to be covered with thin pus. The pieces of wood are removed, the surface is cleansed with an alcoholic solution of corrosive sublimate or ether, and the following powder is introduced into the little depressions made by the pieces of wood :

R, Hydrarg. chlorid. corros.,   0.10 (gr. iss);
    Magnes. carbon.,   10 (ℨiiss);
    Acid. salicyl.,   5 (ℨj gr. xv);
    Cocain. muriat.,   0.50 (gr. viiss).

The surface is then again covered with a plaster.

Schütz, under an anesthetic, removes all soft tissue with the sharp spoon, and very carefully scarifies the floor of the wound and about three-fourths to one centimeter of the surrounding healthy border. The entire wound

is then repeatedly painted with a cold **saturated** alcoholic solution of zinc chlorid, **to** which a little pure
hydrochloric acid has been added to make it keep and
remain clear.   Very severe pain follows this procedure;
the **area** operated upon and the surrounding tissue swell
moderately.   Compresses of boric-acid solution cause the
symptoms **to** disappear gradually, and in one to two days
the wound is clean.   An ointment of pyrogallic acid and
vaselin (1 : 4) is then applied; this should be changed
three times daily.   On the fifth day the ointment is
replaced by compresses of boric-acid solution.   After
the eschar has separated, the parts are again treated with
the pyrogallic-acid ointment, and after a suitable interval,
during which compresses of boric-acid solution are again
applied, the pyrogallic-acid ointment is used for the third
time.   Cicatrization **takes place** under empl. hydrargyri,
iodoform bandage, or boric-acid ointment.

Elsenberg has recommended parachlorphenol as a
caustic.   The other remedies which have been advised,
such as injections of thiosinamin (H. v. Hebra), can-
tharidin (Liebreich), tuberculin (Koch), tuberculocidin
(Klebs), have not stood the test of unbiased criticism.
Experience with the latest suggestions, such as the hot-
air treatment (Hollaender, Lang) and illumination with
X-rays, is not as yet sufficient to warrant an opinion.

B. SCROFULODERMA (TUBERCULOSIS CUTIS COLLI-QUATIVA).

The primary lesion and clinical feature of this disease
is the soft nodule.   This is characterized by colliquation
and formation of a fluctuating tumor.   All the pathologic
processes in scrofuloderma have their starting-point in the
subcutaneous lymphatic glands and channels; and in some
instances even in diseased bone.   Inflammation and new
formation of nodules take place beneath the still movable
skin.   Later the nodular infiltration softens, the overlying
skin is firmly attached and finally broken through, and an

indolent, undermined ulcer results. When the process extends, new **tubercles, fistulæ, ulcerations, and cicatrices form.** Occasionally dispersed tubercles (gommes scrofuleuses of the French) are found on parts of the body where we are not accustomed to meet with lymphatic glands. We have observed numerous abnormally-situated lymphatic glands in syphilitic individuals, and agree with Jadassohn that subcutaneous tuberculous nodules occurring in such localities should be regarded more often as abnormally-situated lymphatic glands.

Such typical cutaneous and subcutaneous tubercles also occur in the course of large lymphatic vessels, and are subsequent to the skin-affection (Plate **47,** *b*), or occur independently of such a condition.

Histologically this tuberculous inflammation is characterized by being more sharply defined than lupus and by the greater abundance of pus-corpuscles containing fragmentary granules. The bacilli are few in number; experimental inoculation, however, succeeds better than that made with lupus-tissue; and the animals experimented upon perish more rapidly of general tuberculosis.

Treatment.—The general health must be looked after with the greatest care and the deteriorated condition of the nutrition must be improved as much as possible. Locally, surgical methods are especially indicated, and the after-treatment is to be conducted on general surgical principles. [In superficial conditions the treatment is essentially the same as in lupus.—ED.]

C. THE TUBERCULOUS ULCER (TUBERCULOSIS ULCEROSA CUTIS).

This form of local tuberculosis, also known as miliary tuberculosis of the skin, is usually associated with grave general tuberculosis, and is due to autoinoculation or to extension from the mucous membranes. It occurs in the cavity of the mouth, on the lips, nostrils, anus, and genitalia. The miliary tubercles (formation

11

of lymphoid cells predominates), the size of a pinhead to that of a hemp-seed, show a great tendency to softening accompanied by destruction of the diseased tissue of the skin. A superficial ulcer with a torpid base results, whose margins are serrated, eaten away, and undermined, and with outlying new lesions at the border. At the periphery depressions may occasionally be seen after the miliary tubercles have disappeared, or small whitish-yellow nodules are present (Plates 47, *a*, 47, *b*, 49, 50, 51). The ulcers, especially on the mucous membranes, show a tendency to papillomatous vegetations.

Numerous bacilli are found in this form.

D. TUBERCULOSIS VERRUCOSA CUTIS.

This form of cutaneous tuberculosis, first described by Riehl and Paltauf, is characterized by warty, papillary outgrowths on the surface and by the absence of ulcers and a dearth of lupus-nodules; pustules, however, often develop. It occurs on the fingers or dorsal surface of the hand, and is found in butchers, attendants in morgues, and in physicians—in brief, in those having to do with manipulation of tuberculous material, and is consequently the result of exogenous inoculation. Post-mortem tubercles, scrofuloderma, or tuberculous ulcers may also result from infection of this kind.

Tuberculosis verrucosa cutis is a localized process. The grayish-white, warty papillomata may appear singly or in groups, and exhibit a tendency to heal in the center and to spread at the periphery. Fully-developed tubercles are found in the most superficial layers of the cutis, and contain bacilli, with coexisting small-cell, diffuse infiltration. The pustules mentioned above are minute miliary abscesses in the small-cell infiltration, associated with collection of pus beneath the epidermis. Cocci have been found in the purulent matter of the pustules, which the authors already mentioned regard as the pyogenic factors. The epidermal involvement, the proliferation of the stratum Malpighii and corium, and extension of the rete, which is traversed by leukocytes, are the result—as is the case in many other diseases—of inflammatory processes in the superficial cutis, in which, as has been referred to, the deeper layers of the epidermis participate, with proliferation and cornification of the upper layers.

Treatment.—Erasion or excision.

E. TUBERCULOSIS FUNGOSA (FUNGUS CUTIS).

Riehl has described a tuberculous infiltration beginning deep in the bone and periosteum and progressing upward toward the soft parts, which leads to formation of fistulous tracts and to soft superficial growths, giving rise to mushroom-like tumors, which disintegrate *de novo* and form ulcers. We have described such a variety on the lower extremity (Plate **47, c**). In such a case it is quite proper to drop the term lupus, inasmuch as in this instance, as in tuberculosis verrucosa, lupus-nodules do not occur, the disease being characterized solely by infiltration and subsequent disintegration, but not by colliquation, as is the case in scrofuloderma.

According to Riehl, bacilli are more numerous in this manifestation than in scrofuloderma or lupus.

LEPRA.

Leprosy (Plates 51, *a, b, c*) is a chronic infectious disease, due to a specific bacillus, and consists of the formation of granulation-tissue growths of varying character and extent.

In Europe it is most common in Norway, the Swedish, Finnish, and Russian coasts on the East Sea; in Asia, in India, China, Africa, Egypt, Abyssinia, Morocco; and in America in California and Mexico, in Australia, and Sandwich Islands.

The cause of leprosy is the Bacillus lepræ (Hansen, Neisser), and its discovery has been the means, contrary to the older views, of adding more believers in the contagious nature of the disease, and that it spreads from individual to individual.

Two chief forms are usually described : *lepra tuberosa* and *lepra anæsthetica seu nervosa*. As the essential cause is the same in both, it can be readily understood that mixed forms are frequently encountered. There are certain sites of predilection on the general surface, although leprous nodules are constantly found in the liver, spleen, lymphatic glands, and scrotum in both forms.

Lepra tuberosa, or **tubercular leprosy,** attacks chiefly the integument and the mucous membranes of the nose, palate, roof of the mouth, larynx, and pharynx.

On the skin the first changes show themselves in the form of infiltration ; the skin in one or more places, over areas of several centimeters, becomes elevated and assumes a brownish-red or dull-red color. In the region of the infiltration the sensibility disappears partly or completely, and on hairy parts the hair of the affected area falls out.

After a longer or shorter period (up to several years) there develop upon these patches nodular and tubercular growths ; they appear as papular lesions, brown to copper-brown in color, and gradually increase in size. In the beginning scattered or discrete, they may later form by confluence diffuse masses with a rough, uneven surface. The size of the lesions or patches may vary between that of a pea and an extensive tumor-like mass. They are hard in consistence, and the skin-sensibility is reduced or abolished. The favorite site for the tuberculous lesions is the face, especially the forehead, eyebrows, nose, and lips ; likewise the upper and lower extremities, especially the extensor aspects. After variable duration the tubercles undergo changes, either becoming fibrous with atrophy, or softening and breaking down. Ulcerations covered with grayish coating and with callous borders are the result of this disintegration. Direct suppuration of the nodules is somewhat rare. The ulcers extend deeply to sinew and bone, the latter being laid bare and necrosing; at times also the joints are in this manner opened up.

On the mucous membranes the lesions show themselves either as small papules or tubercles, or as round, flat infiltrations, which become ulcerated and may heal with cicatricial shrivelling. The results are often conspicuous disturbances of the affected part — disappearance of the cartilaginous nasal septum, the soft palate, and the epiglottis; stenosis of the larynx is one of the most common occurrences.

Also on the conjunctiva bulbi, especially at the corneal border, characteristic tubercles often develop.

The disease has a remarkably regular and progressive course, inasmuch as new lesions are always presenting themselves. The new outbreaks arise, as with the initial eruptions, under febrile action; erythematous reddening of the affected parts presenting, which is soon followed by the formation of tubercles and nodules. At the sites of the older lesions, usually at the time of the fresh outbreaks, changes are noted to take place, miliary abscesses or blebs arising, either of which may end in ulceration.

It is deserving of mention that at the time of these fresh outbreaks the lepra bacillus may be demonstrated in the blood, in which at other times it is wanting.

Lepra Anæsthetica seu Nervosa.—Anesthetic leprosy is characterized by sensibility and trophic disturbances of the skin and muscles, the new tissue-formation, which produces the nodose growths of the tubercular form, remaining in the background or entirely wanting.

The disease begins as a leprous polyneuritis. Its substratum is the leprous deposit, with but slight granulation-tissue formation (leproide) in the peripheral nerves. In the early stage rounded spots appear, often confluent, and for the most part symmetric, of a bright, later dark-red color, which in time changes to a brown or dark brown. The spots grow by peripheral extension to palm size, and usually show a slightly infiltrated edge and an atrophic center. The more recent the eruption the wider the border of infiltration. With increasing atrophy the color becomes paler and paler, changing to a yellowish-brown, the pigment finally disappearing, so that the atrophic areas are then lighter in color than the surrounding skin. The increase in the affected area takes place by the gradual creeping outward of the infiltration, while the inner portion atrophies. Through this manner of spreading and through confluence of neighboring patches map-like areas are produced. The attacked parts are completely anesthetic.

The sensibility and atrophic symptoms are the predominant characteristics of this type of leprosy. Soon follow deep-seated disturbances of sensibility, first thermo-anesthesia, later complete anesthesia of the skin, and finally anesthesia of the deeper parts, muscles, and bones.

Among the atrophic disturbances the first are atrophies of the muscles, with preference for the thenars (Aran-Duchenne type), the interosseals, and the extensor muscles of the hands. On the lower extremities the first muscle to be attacked is usually the extensor of the toes. Later there is noted involvement of other nerve-regions, especially the face.

In addition to those already described, the atrophic disturbances of the skin are ulcer-formations—pressure-ulcers—which are observed most frequently in the form of perforating ulcer of the foot. Further, there appear in the palm deep fissures and rhagades, which may extend to the fingers and to the dorsum of the hand.

One of the most frequent lesions of the skin is bleb-formation, the so-called pemphigus leprosus. The blebs vary in size from a pinhead or pea to a grape or larger, are filled with clear liquid, break, and leave livid excoriated spots, which by neglect or improper treatment may give rise to ulcers. The appearance of the bleb-eruption is usually accompanied by general symptoms. Some investigators have stated that they were able to find lepra-bacilli in the blebs.

The deeper parts also show trophic disturbances; especially are the bones of the finger-phalanges so disposed. These become necrotic, the phalanx swells, softens, and breaks down into a fistule, through which the bone is cast off. The result of this recurrent process is a distortion of the hand, to which the name of *lepra mutilans* is given.

Worthy of mention is the recent conclusion of Sticker, which points to the primary effect of lepra as a specific lesion of the nasal mucous membrane, especially in the form of an ulcer over the cartilaginous part of the septum. From this primary ulcerative lesion lepra-bacilli are being constantly thrown off in enormous numbers.

The **course** of the disease is eminently chronic, the duration extending between five and eighteen years, the anesthetic type being the more prolonged in its course. A cure is unknown; all cases end fatally. [Several alleged cures, or at least apparent cures, have been reported from time to time.—Ed.] There develops a progressive cachexia, due to the persistent ulcerations of the skin and to the severe trophic disturbances, and also to visceral leprous complications (liver, spleen, kidneys).

Especially the kidneys show constantly severe parenchymatous changes, without necessarily being the seat of the leprous deposits. Visceral leprosy induces severe derangements of the stomach and intestinal functions, so that the patient may succumb to the increasing cachexia so caused or to some fatal intercurrent affection. Especially is tuberculosis one of the most frequent complications which carry off the patient or hasten the fatal end.

Treatment.—The greatest weight is to be placed upon prophylactic measures. In cases in which the disease is already established dietetic and hygienic measures play a very important part in its management, without being sufficient to stay materially the progress of the disease. The remedies proposed for the treatment of leprosy, even including the Carrasquilla-serum, have a problematical worth. Unna claims that by the administration of ichthyol and the local application of ichthyol and pyrogallol to have cured two cases. Also the internal use of sodium salicylate and iodid preparations has been praised by many. Vidal gives:

R Balsam. gurguni,
    Acaciæ,   *āā* 4 (ʒj);
    Tinct. catechu,   12 (fʒiij);
    Infus. valerianæ,   60 (fʒxv).
Sig.—For one day.

This daily dose is gradually increased 12 (ʒiij) pro die.

Frequently prescribed is chaulmoogra oil, in the dose of 5 to 120 drops three times daily. Locally:

R Ol. chaulmoogra,   25 (fʒvj gr. xv);
    Vaselini,   50 (ʒxiiss);
    Paraffin.,   10 (ʒiiss).

Or resorcin salve, 5 to 20 per cent.; also ichthyol salves.

MALIGNANT GROWTHS OF THE SKIN.

The general integument is often the seat of malignant new growths, arising spontaneously or through metastasis, the cutaneous manifestation being the first evidence of the disease. Most of these growths belong essentially to the domain of surgery, and are fully treated in works upon that subject. It is, however, often the province of the dermatologist to see these formations in their earlier stages.

The most frequent malignant tumors are the sarcomata and their allied growths, and certain forms of carcinomata.

In the past several years, on both clinical and histologic grounds, many growths heretofore classed under sarcoma have been recognized as distinct formations. Kaposi includes under the name of "sarcoid tumors" granuloma fungoïdes (mycosis fungoïdes), lymphodermia perniciosa, and sarcomatosis cutis, although he recognizes the fact that it is difficult to treat of such differently characterized diseases collectively.

GRANULOMA FUNGOÏDES.

Granuloma fungoïdes is a disease which should be discussed separately from the sarcomata (Alibert). This chronic skin-disease is distinguished by a progressive course and by the formation of infiltrated and tumor-like growths which develop rapidly, but which may also undergo complete involution. Viewed as a whole, it is customary to divide its course into three different stages. The disease begins with prodromal erythematous and eczematous, intensely itchy plaques on the trunk, on the flexors of the extremities, and on the face, especially on the forehead. The epiderm in these places exfoliates or is covered with thick crusts. [This stage may last from several months to several years.—ED.]

Owing to the intense itching, the patient is troubled with loss of sleep. Gradually individual lesions or patches completely disappear, others heal in the central part and spread at the periphery, and there gradually develops what Kœbner has designated the stage of infiltration—second stage. In addition to the infiltrated patches or areas, lentil- to bean-sized red protuberances appear, which gradually develop into half-rounded tumors of the size of a small apple or mandarin orange, and the third stage is entered. The color is pale brown to dark red, the surface notched or serrated, the center slightly depressed. At first hard, it gradually becomes softer. These tumors also may melt away in the course of several days or a few weeks, leaving nothing but pigmentation. More frequently, however, they become necrotic and give place to ulcers which bleed readily. The patients' general condition, apparently little disturbed in the earlier stages, now begins to fall perceptibly; they become marasmic, and the large majority gradually succumb to the disease.

The lymphatic glands are not involved in the process. In exceptional cases, at the autopsy, numerous whitish bean-sized tumors have been found in the internal organs.

Histologic investigations of the tumors of granuloma fungoïdes teach that the process consists of cell-growth about the vessels, at the bases of the papillæ in the connective tissue, and about the glands and hair-follicles. The cell-growth appears mostly as an infiltration crowding out the cutis and the papillary body. The irregular collections of round cells are massed in a framework of fibrillar connective tissue; and Paltauf intimates that this stroma for the most part consists of bundles of cutis-fibers pressed asunder. The epidermis in the beginning seems thickened; later, however, it is thin and free from proliferation-processes.

Unna calls special attention to the fact that parasites can easily localize themselves in the loose, soft tissue, and may easily lead to necrotic changes and general septic infection.

The various findings of bacteria and cocci in the growths are to be looked upon as belonging .to septic processes and accidental, and not necessarily having any pathogenic relationship to the disease.

Most authors are agreed that the tumors of granuloma fungoïdes occupy a middle position between granulation-tumors and sarcomata; in support of such view we have the relatively benign character of the disease, the spontaneous involution, and the slight disposition to metastasis, in addition to the anatomic changes.

In the external treatment the reducing remedies are most commonly employed—resorcin, chrysarobin, and pyrogallic acid.

The best results are promised from arsenical administration, along with the external use of the remedies named. Surgical treatment is without permanent results.

The skin-manifestations in *leukemia* and *pseudoleukemia* consist of various tumors and infiltration-lesions, which, judged by external appearances and form, seem to approach closely to granuloma fungoïdes and also to sarcomata. Paltauf (*Transactions of the Second International Dermatologic Congress*) calls attention to the existence of the general disease before and at the time of the **development of the tumors, and** rightly emphasizes that we are **enabled by the blood-**investigation, especially **in leukemia,** very **early in the course of** the disease **to render a** diagnosis as **to the nature of** the tumors and skin-infiltration. The blood-investigation discloses **a true** leukocytosis. The **number** of the red blood-corpuscles **is more or less** diminished **and** the hemoglobin decreased. Some caution is required, however, inasmuch **as** similar conditions **are** sometimes met with in granuloma fungoïdes **and also in** sarcomatosis cutis.

Lymphodermia perniciosa **Kaposi describes as a disease** characterized **by** eczematous manifestations **and the development of infiltration and** nodes, which **may be seated upon the face, trunk**, and extremities. The spread of **the disease over** the forehead, ears, and lips gives the patient **the appearance of** facies leonina.

The disease appears **as** leukemic tumor-growths **or as diffused** infiltrations in **the** subcutaneous fat-tissue, **over which the** skin **is** eczematous. With increase **of** the **general paleness** individual **growths** break down and change into ulcers; swelling **of the** lymphatic glands also develops, as well **as** enlargement **of the** spleen, the patient finally succumbing. **At the** autopsy leukemic nodes **are** found in the pleura, the lungs, and other internal organs, as well as **in the** skin and glands.

Similar, **if less** characteristic, appear the **deposits in the** skin in *pseudoleukemia*. In this disease also **eczematous or urticarial** manifestations usually go hand **in hand** or precede **the** node-formation in the subcutis. The subjective symptoms, as well as the further course, are similar **to** those of leukemia, which together with the blood-investigation **permit a recognition of the disease.**

SARCOMA CUTIS.

The distinctive sarcomata **of the** skin appear as the **typical** melanotic sarcoma, *sarcoma melanodes.* These tumors arise from a warty growth **or nevus, and within a few** weeks **result in pea-** to cherry-sized, **and also** larger, painful, dark-blue growths. At first they are hard, but later **become** more succulent. **The** lymphatic glands become swollen, the nodes **break** down, and through confluence there arise larger blue-black plaques. Finally metastases take place in the internal organs and the general cachexia leads to a fatal end.

The melanotic sarcomata are alveolar angiosarcomata with pigment-deposit in and between the cells.

Another form is multiple, hemorrhagic, idiopathic sarcoma. Inasmuch as we have had no experience with this form, we give a brief description by Kaposi : " Without known cause hazelnut-sized, bluish, firm and elastic, rounded, elevated, occasionally grouped or bunched, nodes appear having **a** smooth surface, and being at first observed on **the feet.** Later the eruption occurs on the legs, arms, and **trunk ;** and finally swelling **of** the lymphatic glands, and node-formation **in** the mucous membranes and in the internal organs are noted. Individual nodes may undergo involution." The pigmentation Kaposi considered due to capillary hemorrhages. The duration **of** the disease **is from** three **to** eight years, during which **time** new nodes are developing from the peripheral to the central parts. The feet and hands are swollen and painful upon pressure. The involution of the growths, with formation **of** pigmented cicatricial depression, is the usual course with the older nodes ; breaking-down occurs less frequently. With fever, bloody diarrhea, hemoptysis, and marasmus, death finally takes place. At the autopsy are found vascular nodes in the lungs, liver, spleen, and in the muscles **of the** heart, and especially in the large intestine.

The **treatment** of sarcoma is essentially surgical. In pigment-sarcoma arsenical treatment **should** be tried. This is the only method which **so far has** given **a** good result (Kœbner).

EPITHELIOMA (RODENT ULCER, CANCER, SKIN-CANCER).

The general integument may be primarily the seat of epitheliomatous growths, or it may be involved secondarily from tumors beneath the skin ; or skin epithelioma may finally occur as a metastasis from one or more of the internal organs.

The most frequent primitive form on the skin is the epidermic cancer. In the beginning it appears as a flat hard papule or tubercle, or as a diffused, uneven, irregular growth, or as a subcutaneous nodule involving the skin. The chief characteristic of this form of epithelial cancer is the so-called pearl-rolls or bodies, the cancroidal bodies, which appear as a conglomeration of variously-shaped epithelioid cells in the form of waxy, glistening or pale-red hard tubercles, which if seated on the surface may be readily pressed out. For several years or more a flat wart-like growth presents, newer nodules forming on the periphery. If the mass breaks down, a flat super-ficial ulcer (ulcus rodens) results, secreting scanty fluid, which dries to a thin covering or crust.

Sometimes there results complete exfoliation with cica-tricial formation in the center, a new progressive hard, waxy-looking edge with contained cancroidal bodies form-ing on the borders. Should the scar and the border contain pigment, it represents the so-called chimney-sweepers' cancer.

For ten to twenty years such a process upon the skin may go on, apparently at times stationary ; sooner or later induration, ulceration, contractions, and consequent changes in the skin take place, but without the general organism being disturbed.

Some epitheliomata arise out of nodular, more deeply-seated tumors, which may reach down to the subcutaneous tissue, forming flat growths which break down in the central part. These break down earlier than the type first de-scribed ; also spread into the peripheral region more quickly; may, however, cicatrize in the center, so that surrounding a shiny vascular scar-tissue is noted a garland of fresh epithelial formation or tissue. The growth may also be papillomatous, which breaks down more quickly, following the course of the more malignant form of this variety of cancer.

The most frequent sites of epithelioma are the eyelids, nose, lips, and less frequently the forehead and cheeks. Of importance are the epitheliomas of the eyelids, which gradually destroy the latter, invade the conjunctiva, and finally the bulbus (Plate 55, a).

From the nose and lips the epitheliomatous growth may extend to the mucous membrane of these parts. The disease may also occur primarily as an independent affection on the mucous membranes of the mouth, nose, and rectum. The frequent thickenings observed on the mucous membrane of the cheeks, and especially the tongue, are after years' duration often the starting-point of epithelioma. On the penis, especially about the urethra, epithelioma develops, and invades the corpus cavernosum, forming small or large ulcers (Plates 54 and 55). The lymphatic vessels of the penis and the inguinal glands become involved; at first hard painless tumors form, which may break down and become purulent. Epithelioma of the external genitalia and vagina of women behaves the same way, and may frequently be mistaken for syphilis (Plate 53).

Epithelioma occurs generally in advanced years. It may appear at the site of slow granulating ulcers or scars after syphilis and lupus; or have its seat, as already mentioned, in warts and mucous-membrane thickenings. It may exist, as already indicated, ten to twenty years without endangering life, till finally, more especially in the papillomatous form, more rapid breaking down and glandular involvement ensue and the patient dies from marasmus.

Another form of cancer observed in the skin is *carcinoma lenticulare*, which frequently starts from mammary

cancer, with redness and hardening; spreads and gives
rise to an infiltration of the skin, so that the thorax is
covered with newly-formed masses, as if enveloped in a
coat of mail (cancer en cuirasse) (Plate 52).

Treatment.—Surgical methods are of first importance.
Only when surgical treatment cannot be carried out is
recourse to be had to other plans. As such, we name the
destruction of the growth with caustics (lactic acid, acetic
acid, nitric acid, Vienna paste, zinc chlorid, arsenical
pastes), thermocautery, and erasion of the mass with the
curet. The pyoktanin treatment, as likewise the Adam-
kiewicz's cancroin treatment, has been abandoned. In
ulcus rodens, resorcin, pyrogallic acid, in powder or salve
form (15 per cent. to 30 per cent.), has been recommended. Lassar recommends subcutaneous arsenical injections. [Many of these cases, and especially in the
early stages, and those of a superficial type, can be most
satisfactorily treated with arsenical and zinc-chlorid
plasters.—ED.]

PARASITIC DISEASES OF THE SKIN.

The parasites of the skin are of both vegetable and
animal nature. The diseases induced by their presence
have naturally a contagious character; such diseases are,
however, distinct from infectious diseases, which are also
called forth by parasites (microörganisms), but which, in
addition to attacking the skin, involve other organs.

The vegetable parasites of the skin belong collectively
in the group of pathogenic mould-fungi (hyphomycetes).

The diseases produced by these parasites are termed
dermatohyphomycoses or dermatomycoses.

Each of the several disease is produced by a special
fungus.

TINEA FAVOSA (FAVUS).

Favus (Plate 56) is due to invasion of the skin by a
vegetable parasite, the Achorion Schönleinii. This fungus

consists of numerous wide and branching mycelial threads and spores, **is found** usually on the hairy scalp, and forms **disc-like** yellowish **crusts,** which show in the center a depression. **The fungus invades the** follicles and even the sheath **of** the hair-root, and causes falling of the hair. The hairs of the affected **spots** may be pulled out easily or readily break off. The **color** of the discs or crusts is sulphuror straw-yellow. **After the** crust is removed or falls off there is left a smooth atrophic depression. The follicles **are** destroyed and the affected areas are more or less bald. Through confluence large masses of crusts are **formed.** The fungus (Plate 65, *e*) gives out a characteristic mouse-urine odor.

This chronic disease, which usually appears early in life and persists through adolescence and manhood, may disappear spontaneously, all the follicles having been destroyed. In such cases the scalp is completely bald, with **the** exception of scattered single hairs or tufts of hair; the skin is thinned, smooth, **and atrophic.**

The disease **is** also met **with on other parts of the** body, although comparatively seldom. The fungus has, in fact, **been** found, in a case of universal favus, even on the mucous membranes, the patient having died of gastro-enteritis (Kundrat). The nails of the hands may also be the seat **of** this vegetable parasite, with the consequent changes; **they become** opaque, crumble, **or** break easily, and are found permeated with the fungus (Onychomycosis favosa).

Treatment.—Treatment **of** the disease on the scalp begins with cropping **short the hair of** the whole region. After this **the** accumulated fungus-masses are removed. This **is** most readily accomplished by softening with oil **or** fats, with **or** without the **addition of** carbolic acid or naphthol; and subsequently by thoroughly shampooing. When this **has been** effected **the** diseased areas should be depilated, **and this** should not **be** limited **to** the spots, but should **extend one or two centimeters** beyond the borders. **By** gentle **traction** only **the** diseased hairs are brought away. Lotions of antiseptic solutions and bandages spread with antiparasitic salves may be applied, having in view the destruction of the fungus.

The number of such applications for this disease is a very large one. We name the tar preparations, salicylic-sulphur salves, alcoholic solutions of corrosive sublimate, resorcin, naphthol, creolin, pyrogallic acid, and chrysarobin :

℞ Chrysarobin,
  Ichthyol,   āā 5 (gr. lxxv) ;
  Acidi salicylici,   3 (gr. xlv) ;
  Vaselini,   100 (℥iij).—M.
Ft. unguentum (Unna).

Besnier advises the following salve to be applied at night :

℞ Bals. peruviani,
  Acidi salicylici,
  Resorcini,   āā 5 (gr. lxxv) ;
  Sulphur. præcip.,   15 (℥ss) ;
  Lanolini,
  Vaselini,
  Adipis lanæ,   āā 30 (℥j).—M.
Ft. unguentum.

In the morning the scalp is thoroughly washed with lukewarm water and soap (tar-naphthol soap), dried, and then the following solution painted on :

℞ Spirit. vini gallici,   100 (℥iij) ;
  Acidi acetici,   0.25–1 (♏iv–♏xv) ;
  Acidi borici,   2 (gr. xxx) ;
  Chloroformi,   2 (♏xxx).—M.
Sig.—External use.

Pick considers the best method of treatment to consist of daily washing with boric-acid soap, and subsequently applying a 5 per cent. to 10 per cent. alcoholic solution

12

of boric acid; in severe cases powdering with boric-acid powder, over which is placed moist lint, and then enveloping the parts with gauze.

Pirogoff orders the affected parts shaved, and every twenty-four hours the following salve to be applied, spread as a plaster:

℞ Potass. carbonat.,	8 (ʒij);
Sulphur. sublimat.,	30 (ʒj);
Tinct. iodini,	
Picis liq.,	āā 100 (ʒiij);
Adipis benzionat.,	200 (ʒvj).—M.
Ft. unguentum.	

Each time before the salve is applied the scalp is to be washed with soap and water.

Zinsser orders the scalp washed with soap and water, and shaved; the scalp is then covered with compresses wet with a solution of 3 per cent. carbolic acid or of 0.25 per cent. corrosive sublimate, over which is placed a Leiter coil, through which water of the temperature of 52° to 58° C. is kept circulating. During the night the coil is not employed.

In carrying out any of the plans mentioned above for the treatment of this obstinate disease persistence must be enjoined for many months. Culture-tests of the depilated hairs must be made the basis of further treatment or its discontinuance.

The treatment of favus on non-hairy surfaces is much easier and more satisfactory. The crusts are removed, and one of the antimycotic applications already mentioned applied to the affected area.

The treatment of favus of the nails consists in bathing the parts in antiseptic solutions, and then applying compresses wet with the solution. Before making the application the nail should be thoroughly scraped with the sharp spoon or gently cut away.

TINEA TRICHOPHYTINA (TRICHOPHYTOSIS, RINGWORM, HERPES TONSURANS (OF THE GERMANS)).

The several cutaneous manifestations due to invasion of
the cutaneous tissues by the trichophyton fungus (Gruby,
L. Malmsten), usually designated tinea circinata, tinea
tonsurans, and tinea sycosis, present externally diverse
appearances. This vegetable-parasite, consisting of long
mycelial threads with comparatively few spores, vegetates
in the upper layers of the epidermis and gives rise to
greater changes and more diverse clinical pictures than
does favus. The upper layers of the skin become slightly
or moderately inflamed, with scaliness and vesicle- or even
pustule-formation. [It seems now to be established that
there are two distinct forms of fungus responsible for ringworm—the small-spored fungus (Microsporon Audonini)
and the large-spored fungus (Trichophyton). Of the latter
there are several varieties.—ED.]

TINEA CIRCINATA (TINEA TRICHOPHYTINA CORPORIS).

In average cases of tinea circinata—ringworm of non-
hairy parts—one or several pinhead- to pea-sized slightly
hyperemic spots appear, which soon show slight branny
scaliness; the central part begins to clear up, while the
patch enlarges by spreading peripherally. After several
days or a week they usually attain the size of a silver
quarter. The border is noted to be slightly red and
scaly, and may even tend to papular and vesicular formation, or in exceptional cases small pustules may develop.
The central part clears up, the skin being there pale red
or pale brownish, free from scaliness or with trifling
exfoliation. The outer part of the circle is usually somewhat more scaly, but this is rarely pronounced. The disease may remain stationary, or the patches may extend
somewhat; or new spots may show themselves. As commonly met with there are rarely more than three to ten

areas. **The older** patches gradually disappear **with slight scaliness.** This frequently takes place after one **or two weeks;** usually **as** the result of the application of some home-remedy **of an** antiseptic character, or **it may** spontaneously disappear. In some **cases the areas** are persistent and demand **more** energetic **applications,** which will be **referred** to later.

[Under the name " herpes tonsurans disseminatus," the author describes a manifestation, which is considered in this country to be independent **of** the ringworm-fungus, and **to** represent the disease known as *pityriasis maculata et circinata.* At all events, **it** represents in its clinical manifestations the disease here referred to, and the **atlasplate** (Plate 57), which **in** the original is put down **as** illustrative of " herpes maculosus et squamosus," has accordingly been changed to that **of *pityriasis** maculata et circinata.* The author's description, somewhat abbreviated, will be given **in his words and with** his title.—Ed.] " Herpes tonsurans **disseminatus** [pityriasis maculata et circinata—Ed.] presents itself over **extended** surfaces **(abdomen, back,** breast) in rapidly successive, small **pale-red spots with** irregular borders, which present in the **center** a small **scale.** Near by, and especially on the lower parts, new spots develop **in** a few days. The older scales in the center extend irregularly toward the peripheral parts, **so** that **the** center **may** have entirely recovered and the scaliness **be** found chiefly on the outer portions. Sometimes before this general outbreak **an** old circumscribed patch may be found. The patches often attain the size of coins. Owing **to** the peripheral spread and **the** central involution they are often annular, **the** central part is finally without scaliness and merely pigmented, the peripheral part still scaly, reddened, and covered with flat adherent scales."

Eczema marginatum (**Plate** 26) is **a** name originally given to **a** disease involving usually the crurogenital region, which **was** subsequently found **to** be due **to** the ringworm-fungus. **It** arises on sweating, superficially-macerated regions, which furnish **a** good soil for the vegetation of **the** parasite. The skin becomes infiltrated, reddened, **and** scaly, **and shows** peripherally a sharply-**defined**, elevated edge **beset** with vesicles and crusts. **The nature** of **the** region involved prevent the involu-**tion** which takes place in patches of **the** disease when seated elsewhere ; instead, the skin thickens, and is either reddened or pigmented. Through confluence of several such areas the disease may involve **the** whole **genital** region, thighs, scrotum, and extend upward beyond the pubes ; it is irregular in outline, and gradually spreads **out**-ward. This disease, owing to heat, moisture, and friction of the parts, is very troublesome, itchy, and painful ; especially in soldiers after long marches.

In a similar manner to **that** just described the regions of the axillæ, the anal fold, and the under part of large loose-hanging breasts in women may be the **seat of** the **disease.**

In ringworm, as in favus, the nails may **also** be in-volved, together with the disease on other parts or inde-pendently (onychomycosis trichophytina). The fungus presses into the nail-substance, and it may in this way become opaque in spots or the entire nail may become milky and fragile. Less frequently the nails may **be** more severely involved—increased in size, bent, and dis-torted (onychogryphosis trichophytina). It is extremely persistent, much more so than ringworm of non-hairy parts, and may even **be** more so than **the** disease upon the scalp.

Tinea **Tonsurans** (Tinea Trichophytina **Capitis**).

Tinea tonsurans, or ringworm of the scalp, presents **at** first a somewhat similar appearance to **a** patch **of** the dis-ease on other parts. These characters are, however, soon **lost.** The fungus penetrates the hair-substance, between the cells of the cortical substance ; the hairs become lusterless, break easily, and some fall out. The broken ends show brush-like extremities. Some break off just at the margin of the follicle and appear as black specks in the duct-opening. The follicular outlets in the earlier stages are somewhat more prominent, like goose-flesh, from the crowding of cells and fungus. One or several patches may be present, and may attain the size of coins or larger; if two or three are close together, they may fuse and an irregularly-shaped area result.

The patches vary in size, and are usually covered with slight scaliness and occasionally with crusting. The fungus tends to press into the hair-follicles, and there may develop follicular and perifollicular irritation, with suppuration and marked exudation; in some cases with considerable circumscribed swelling (tinea kerion). The disease shows no disposition toward spontaneous recovery, though it may remain stationary. [It rarely persists beyond the age of fifteen years, and is only exceptionally met with in the adult.—ED.]

Tinea sycosis, parasitic sycosis, or barber's itch, is a disease of the bearded parts of the face due to the ringworm-fungus. The process may remain a superficial one, resembling somewhat ringworm of the scalp; but more commonly it develops into the classical type of the disease, consisting of considerable lumpiness and nodulation, with more or less hair-loss and suppuration.

The trichophyton is conveyed from man to man; frequently, however, from domestic animals to man, as, for example, from cats and dogs to children, from horses and cattle to those whose occupation brings them in contact with such. Shaving also offers a good opportunity for conveyance of the disease.

Treatment.—In the treatment of ringworm of non-hairy parts all remedies capable of bringing about active exfoliation of the epidermis are useful. The most important of this group is sapo viridis, which is to be applied to the affected areas as a salve, repeatedly rubbed in and permitted to remain till mild exfoliation is set up. A combination with naphthol is commended by many dermatologists; but, according to our experience, it does not seem to be more efficient than the soap alone. Applications of tar, chrysarobin (as salve or 5 per cent. chrysarobin solution in liquor gutta perchæ), corrosive sublimate (1–2 per cent. strength), and iodin tincture are also valuable.

In treatment of the disease upon the scalp, after removal of the crusts or scales in the ordinary manner (see Favus) the hairs of the affected areas are to be extracted, and then one of the antimycotic remedies applied. In general, in addition to those already named, the same remedies employed in the treatment of favus of the scalp may also be used in this disease. Kaposi recommends :

℞ Ol. rusci,	15 (f ℥ss) ;
Sulphur. præcip.,	10 (ʒiiss) ;
Tinct. saponis viridis,	25 (f ʒvj) ;
Spirit. lavandulæ,	0.5 (ℏ viij) ;
Bals. peruviani,	1.5 (gr. xx) ;
Naphtholi,	0.5 (gr. viij).—M.

Sig.—External use.

In ringworm of the bearded region it is also necessary that careful depilation should be practised. The remedies to be employed here, as salves, are chrysarobin (with caution), anthrarobin (10–20 per cent.), resorcin, precipitated sulphur; corrosive sublimate (in solution), gray plaster, iodin tincture, and acetic acid :

| ℞ Acidi acetici, | 10 (ʒiiss) ; |
| Sulphur. præcip., | 2.5 (gr. xxxv).—M. |

Ft. pasta. (Kaposi).

The treatment of ringworm involving the nails is the same as that employed in favus of these parts.

TINEA VERSICOLOR.

Tinea versicolor, pityriasis versicolor, chromophytosis, or, as popularly believed, "liver-spots," is due to invasion of the epidermic tissue by a vegetable parasite, the

Microsporon furfur (Eichstedt). This fungus is readily
recognized under the microscope by the bunching of
large masses of spores with mycelial threads between
(Plate 65, **Fig.** *f*). The fungus invades the outer skin ;
the hairs and nails are not involved. It is to be found
especially in the uppermost layers of the epidermis.
With the exception of the face, hands, and feet, the eruption may be found upon any part of the body. As a rule,
its chief seat is on the trunk, and especially the upper
part, particularly on the anterior aspect. It is practically
never seen elsewhere except in connection with the disease on this region. The lower trunk, the axillæ, flexors
of the arms, the crural fold, and the poplitea are sometimes involved. [In several instances the lower part of
the face has also been invaded, extending from the neck.
—Ed.]

The eruption consists of variously-sized yellowish,
brownish, or fawn-colored spots, not elevated, or at least
not perceptibly so. They may become confluent and
form large irregular areas ; even the whole upper trunk
may be uniformly covered. There is usually slight
branny scaliness, visible upon close examination. The
disease begins with one or several spots, and then gradually spreads and increases. It is usually slow in its
progress, and lasts for years, practically showing little if
any tendency to spontaneous disappearance. In sensitive
skins, especially in women, the eruption may have a pale-
red tint. It gives rise to no discomfort, except slight
itching when the patient is heated, although exceptionally
itching may be quite a factor.

The transference of the fungus has been proved ; but it
apparently requires a peculiar susceptibility of the individual. It is found frequently in phthisical patients ;
and such persons, as well as others affected, are frequently subject to recurrences. After long continuance
the disease may finally disappear in advanced years.

Treatment.—Soap-and-water baths ; applications of
tar, chrysarobin, naphthol, iodin tincture. Wolff recom-

mends alkaline baths, and after the baths the rubbing in of an ointment containing corrosive sublimate, one-fourth to one grain to the ounce.

ERYTHRASMA.

Erythrasma occurs from invasion of the cutaneous tissue by the Microsporon minutissimum (Burchhardt, v. Bärensprung). This fungus permeates the epidermis, and consists of numerous fine threads and conidia; Dr. Reale (Clinic of de Amicis) has succeeded in making cultures. The disease is seen especially where two surfaces come together, as on the inner surfaces of the thighs, in the axillæ, etc.; and is characterized by slightly-scaly, palm-sized, brownish spots. The skin in the involved regions is often macerated, presenting intertrigo. The affection runs a very chronic course.

Treatment.—For treatment the reducing remedies are recommended, as tar, chrysarobin, anthrarobin, pyrogallic acid, or combinations of tar and pyrogallic acid and of tar and naphthol.

ACTINOMYCOSIS.

This disease (Plate 61) occurs most frequently primarily on the jaw or neck. It spreads gradually and gives rise to inflammatory symptoms, infiltration, abscesses, and fistules. In the deeper parts the disease spreads as proliferating granulation-tissue, and may even involve the bones.

The cause of the disease is the ray-fungus (Plate 65, *a*), actinomyces; this fungus is also found in cattle and swine.

It is probable that the assumption that the disease is conveyed to man through vegetable food is correct.

The duration of the process depends somewhat upon its location; generally, however, long-continued suppuration and fever lead to marasmus.

If early recognized, the disease may be limited by energetic cauterization or by surgical measures (thermocautery). Potassium iodid has been recommended for internal administration.

THE ANIMAL PARASITES OF THE SKIN.

The animal parasites of the human skin may be conveniently divided into three classes:

1. Those which live in the skin or subcutaneous tissue;

2. Those which persistently or temporarily live on the skin and suck blood;

3. Those which only accidentally are found upon the skin, and give rise to symptoms of cutaneous irritation.

SCABIES.

In the first class of greatest importance is the *Acarus seu sarcoptes hominins* (Plate 64, Figs. *e, f, g, h*), the cause of scabies or itch, an affection of the skin attended with intense itching. The impregnated female mite penetrates the upper layers of the epidermis and makes a burrow in which she deposits her eggs. After the larvæ have been hatched out they begin to burrow also, and the irritation thus provoked gives rise to irritation of the skin, increased by the uncontrollable scratching, and to various inflammatory lesions of the skin (Plates 62 and **62**, *a*), such as papules, vesicles, pustules, ecthymata, and excoriations.

Treatment.—Thorough application of **one of** the salves to be mentioned, with special care for those parts of the body which are most favored by the acarus, as between the fingers, hands, elbows, axillæ, shoulder-region, breast-nipples, the waist-region, lower abdomen, genitalia (especially in **men**), nates, knee-region, and ankles. After the rubbing the patient is enveloped in a woollen cover or puts on woollen underwear. As a rule, this rubbing is repeated morning and evening for two days, and on the fourth day a bath is to be taken. The patient's bed is to be carefully looked after, and disinfected. For inunctions the following are recommended:

℞ Naphtholi,   15 (ʒiiiss);
   Cretæ alb.,   10 (ʒiiss);
   Saponis **viridis**,   50 (ʒxiiss);
   Adipis benzoinat.,   **100** (ʒiij).—**M.**
Ft. unguentum (Kaposi).

Wilkinson's ointment, **as modified by Hebra**:

℞ Sulphur. sublimat.,
   Ol. fagi,
   Saponis **viridis**,
   Adipis benzoinat.,   *āā* 80 (ʒiiss);
   Cretæ alb.,   5 (gr. lxxv).—**M.**
Ft. unguentum.

Or the salve recommended by Weinberg:

℞ Sulphur. sublimat.,
   Styracis liq.,
   Cretæ alb.,   *āā* 20 (ʒv);
   Saponis viridis,
   Adipis benzoinat.,   *āā* 40 (ʒx).—**M.**
Ft. unguentum.

Or Paltauf's **styrax mixture** (styracis, **4 parts**; ol. olivæ, **1 part**).
Or Peruvian **balsam, about nine grams** (ʒij) **for each** inunction.
Or

℞ Potass. carbonat.,   25 (ʒvj);
   Sulphur. præcip.,   75 (ʒxviij);
   Ol. lavandulæ,
   Ol. caryophylli,   *āā* 1 (gr. xv);
   Adipis benzoinat.,   **q. s. ft.** unguent.

Or a **5 to 10 per cent.** losophan **salve**:

℞ Losophani,   5–10 (gr. lxxv–ʒiiss);
     Leni calore solut. in
   Ol. olivæ,   20 (ʒv);
   Adipis benzoinat., **q. s. ad 100** (ʒiij).—**M.**
Ft. unguentum.

Hardy's method for rapid cure is as follows: The patient is thoroughly and vigorously rubbed with sapo viridis over the entire surface, after which he takes a lukewarm bath. After the bath he is rubbed with Hardy's modification of Helmrich's ointment:

℞ Sulphur. sublimat.,	20 (ʒv);
Potassii carbonat.,	10 (ʒiiss);
Adipis benzoinat.,	80 (ʒiiss).

The salve is permitted to remain on for twenty-four hours, and then the patient again takes a bath.

The irritation brought about by the use of these active remedial applications, as well as that which has resulted from the scabietic irritation itself, is to be treated according to the rules governing the treatment of eczema.

The hair-follicle mite, the *Acarus folliculorum (Demodex folliculorum)* (Plate 64, Fig. *i*), is a harmless parasite, which is observed frequently in acne-cases in the glandular ducts and sebaceous glands, but provokes no irritation worthy of mention.

Cysticercus **Cellulosæ Cutis.** — The larvæ of Tænia solium, the **Cysticercus cellulosæ,** live in pigs, deer, dogs; and also in man, acquired by swallowing the embryos. It is to be found most frequently in the eye and brain, but also in other organs, as well as in the subcutaneous tissue, giving rise to an oval nodule. In the connective tissue the growth reaches about the size of a pea, and causes no discomfort. Owing to its seat in internal organs, however, the disease is dangerous. The cysticercus seldom dies spontaneously; in such event the nodule slowly undergoes calcification.

A tropical parasite, the *Filaria medinensis*, the guineaworm, is to be found in the subcutaneous tissue, especially in the neighborhood of the ankle-joint, etc. The larvæ probably gain access through drinking-water. The painful cutaneous symptoms are accompanied by fever. There arises often a painful tumor or ulcer in which the worm is to be **found**; the disease may, however, disappear without these occurrences under the skin.

Pulex penetrans, the sand-flea, comes from South America. It bores into the skin, especially the lower leg and toes, where inflammatory lesions with pus-formation, **and** even lymphangitis and necrosis, may be caused.

CREEPING-DISEASE.

By this name Crocker has **designated a peculiar** skin-affection which occurs **most** frequently in children, **or** upon exposed situations **in** adults. In Vienna Professor Neumann, and subsequently **Dr. Ehrmann** and Dr. Rille, and Russian and other English physicians, have **also** observed it. It appears as **an** itching or burning **spot, from** which a fine red elevated line extends through **the** skin in any direction. This line is either straight, zigzag, or bowed, quite irregular, and lengthens from day to day. **The** fresh progressive line is bright red, about **1 mm.** wide, and slightly elevated; **the** older **lines are flat and** pale brown. The progress is not **constant, but** limits itself **to** a few hours daily, especially **in the night.**

It **is** believed **that** an animal parasite which **bores** similarly **to a** mole is the cause. Efforts to **secure** the same have up **to** the present time been fruitless.

Therapeutically **it** is advised that the progressive end of the line be excised ; according to experience, it is necessary that considerable surrounding tissue be included.

PEDICULOSIS.

To the second class of animal parasites belong the pediculi **or** lice : the Pediculus capitis (Plate 64, *b*), Pediculus vestimenti seu corporis (Plate **64,** *c*), and Pediculus pubis (Plate 64, **d**). The bite of the louse is attended with intense itching, which causes scratching **and as a** further consequence possibly eczema, as we have already **intimated.**

Pediculosis Capitis seu Capillitii.—In head-lousiness, in consequence of the exudation and eczematous irritation produced on the surface, tangling and matting of the hair, and even plica, may result. The scalp of such a person has a mouldy smell, and only after removal of the hair, which requires much care and trouble, can the scalp with its enveloping crusts be seen. Lice, nits (ova), attached to the hairs are to be found, and even maggots may be present, and complete the picture in cases of gross neglect. [Fortunately such extreme cases are rarely, if ever, seen in this country.—ED.]

Pediculosis Corporis seu Vestimenti.—In pediculosis corporis numerous linear scratch-marks may be seen upon the skin ; in neglected cases eczema, furuncles, and cutaneous abscesses may be produced.

Pediculosis Pubis.—The Pediculus pubis, or crab-louse, is to be found on all hair-regions except the scalp, although its common habitat is the pubic region. The eggs, or nits, as with those on the scalp, are found glued to the hair-shaft (Plate 64, *a*). By careful inspection the lice may be discovered close to the skin at the root of the hair.

In addition to the artificial eczema produced by the irritation and scratching, we not infrequently notice on the trunk and also on the thigh bluish rounded spots (maculæ cæruleæ, Plate 63), which, according to Mallet, are said to arise from injection of secretion from the salivary glands near the middle part of the breast ; these marks disclose the migration of the crab-louse over the skin.

The bed-bug (*Cimex lectularius*) lives in the crevices of furniture, especially beds, and during the night feeds upon man. Its bite gives rise to papules or wheals. Similar lesions are provoked by the *Pulex irritans*, common flea, although the central hemorrhage is more distinct. The female lays its eggs in the clefts of floors and furniture and in dusty places. Such eggs have even been found on the body of dirty individuals.

To the third class of animal parasites belong the harvest-mite (*Leptus autumnalis*) and the tick (*Ixodes ricinus*), which bite into the upper skin, and give rise to papules, wheals, slight edema, and pain. A similar parasite is *Dermanyssus avium,* which chiefly attacks fowl, but may also attack man.

In this class also belong the several kinds of gnats (*Culicidæ*) and flies (*Stomonyidæ*), which suck blood and provoke wheals and other symptoms of irritation.

Also many *Œstridæ* (*Myiasis dermatosa œstrosa,* O. Nagel) are to be found, chiefly in tropical countries, on the skin of man and cause boil-formation.

PLATE 1.

Pompholyx.

S. J., aged **25** years, a laborer, was admitted Jan. 16, 1897. The patient had sought hospital-treatment for the relief of swelling and tenderness of the feet. Sweating of the feet had existed in a mild degree since early childhood. He had previously been an inmate of the hospital in 1894, with articular rheumatism; and at that time the soles of the feet were already the seat of numerous disseminated and confluent plaques of loose epidermic scales and some small vesicles; the nails were thickened and brittle.

Status Præsens.—The malleolar regions are swollen and tender upon pressure. The soles are covered with sweat and are studded with pinhead-sized red papules, persistent under pressure; similar lesions are seen at the edges of the soles, less abundantly on the dorsum of the feet and lower instep. In many places these lesions have changed into vesicles; and in other places, especially the plantar region, these vesicles have become confluent and form large blebs with milky contents. The skin of the entire plantar surface, the borders, and dorsum of the feet, is red, as if inflamed.

Jan. 25.—The vesicles, for the most part, have become confluent and form larger lesions, so that both plantar regions are covered with lentil- to bean-sized milky blebs. The borders of the soles show numerous minute hard epidermic granules, which are seated in the glandular outlets and which can be readily pressed out. Under the uplifted epidermic flakes there is apparently slight depression covered with new epidermis having distinctly visible gland-ducts. In some blebs the secretion has become white, thick, and cheesy. The epidermis between the plaques and more active spots is beset with numerous minute, hard, deep-lying granules having a yellowish aspect. The epidermis of the soles is swollen, sodden-looking, and whitish, and in places reddened as if the result of maceration. The palms are moist.

Examination of the cheesy contents of the blebs mentioned showed epithelium, epidermic flakes, and débris.

The patient was, after a month's treatment with mild and softening salves, discharged; the parts had become covered with new epidermis.

[In the German edition the author describes this plate under the heading "hyperidrosis of the feet with vesicle- and bleb-formation." It pictures to the English and American mind, however, what is usually considered pompholyx; although this latter is rarely limited to the feet, as in this instance. In the text, however, the author refers to this plate when describing pompholyx.—ED.]

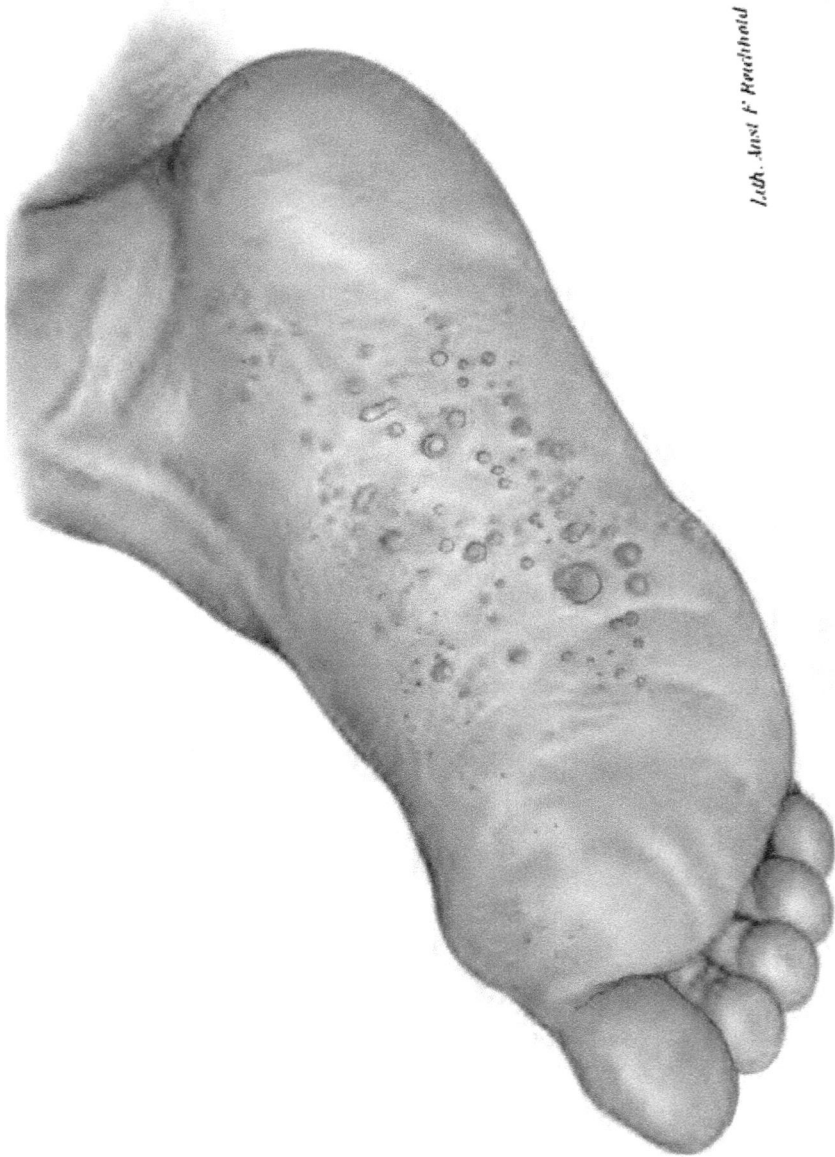

PLATE 2.

Milium.

I. J., aged 18 years, servant-girl, came under notice July 23, 1897.

Status Præsens.—On the face are to be seen grayish and yellowish-white, hard, irregularly scattered, pinhead-sized elevations. By puncturing the overlying epiderm the contents, consisting of firm white bodies, may be readily scratched or pressed out.

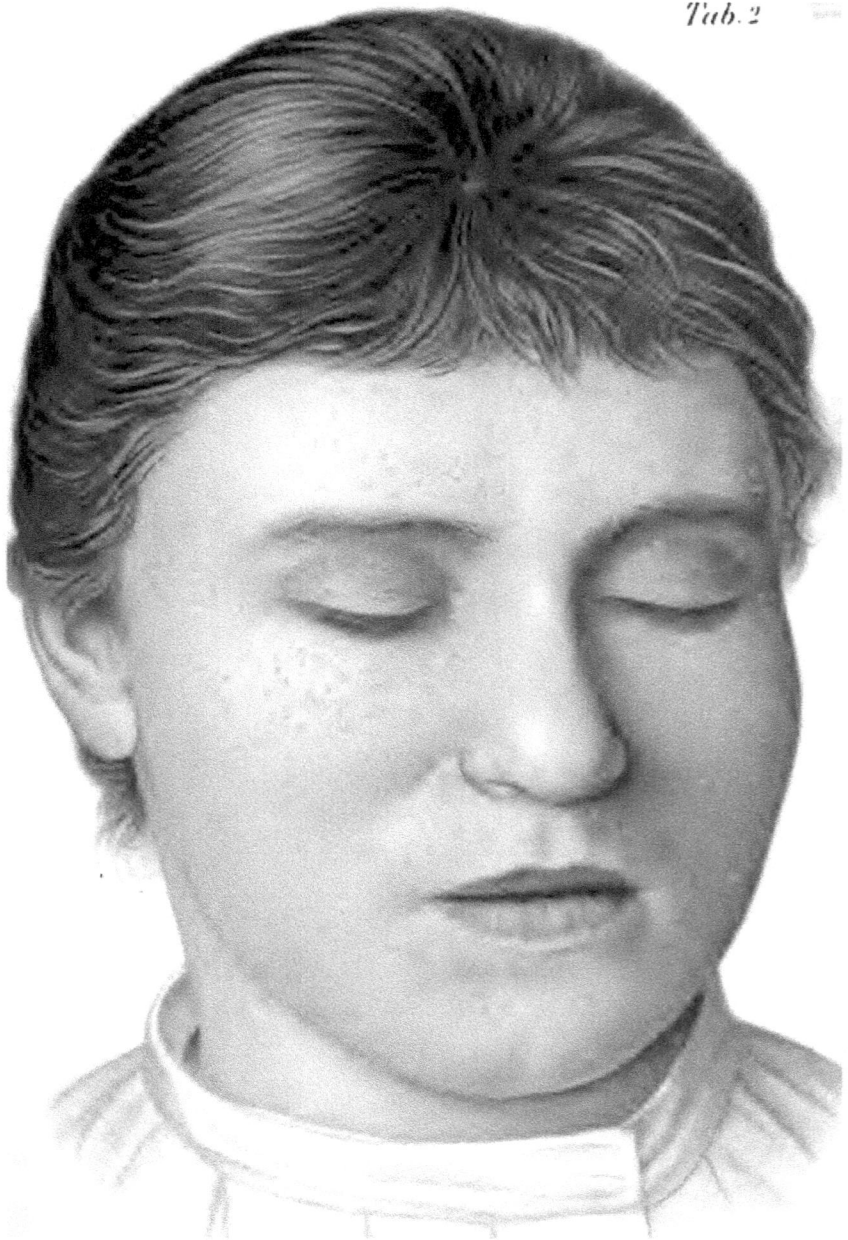

Tab. 2

PLATE 3.

Adenoma Sebaceum. Comedo. Acne.

F. G., aged 22 years, workman, admitted Feb. 18, 1896, states that when 17 years old the inflammatory acne-nodules first appeared; at this time he also noticed the appearance of black points, and the nodular tumors, lentil to pea in size. The disease had now lasted five years.

Status Præsens.—The man is well developed, pale, with moderate amount of flabby panniculus adiposus. The extensor surfaces of the extremities show lichen pilaris. On the forehead, alæ of the nose, especially in the nasolabial folds, on the cheeks, more particularly toward the scantily-bearded portion, numerous comedones are to be seen; scars from former suppurating follicles, acne-nodules, and adenoma in the region of the chin. In the clavicular region are sparsely-scattered comedones—in great numbers, however, over the sternum; also adenomata, and scars varying in size from a pea to a dime, resulting from similar previous growths which had suppurated.

The back is thickly beset with acne-lesions, brown pigment-spots, and comedones

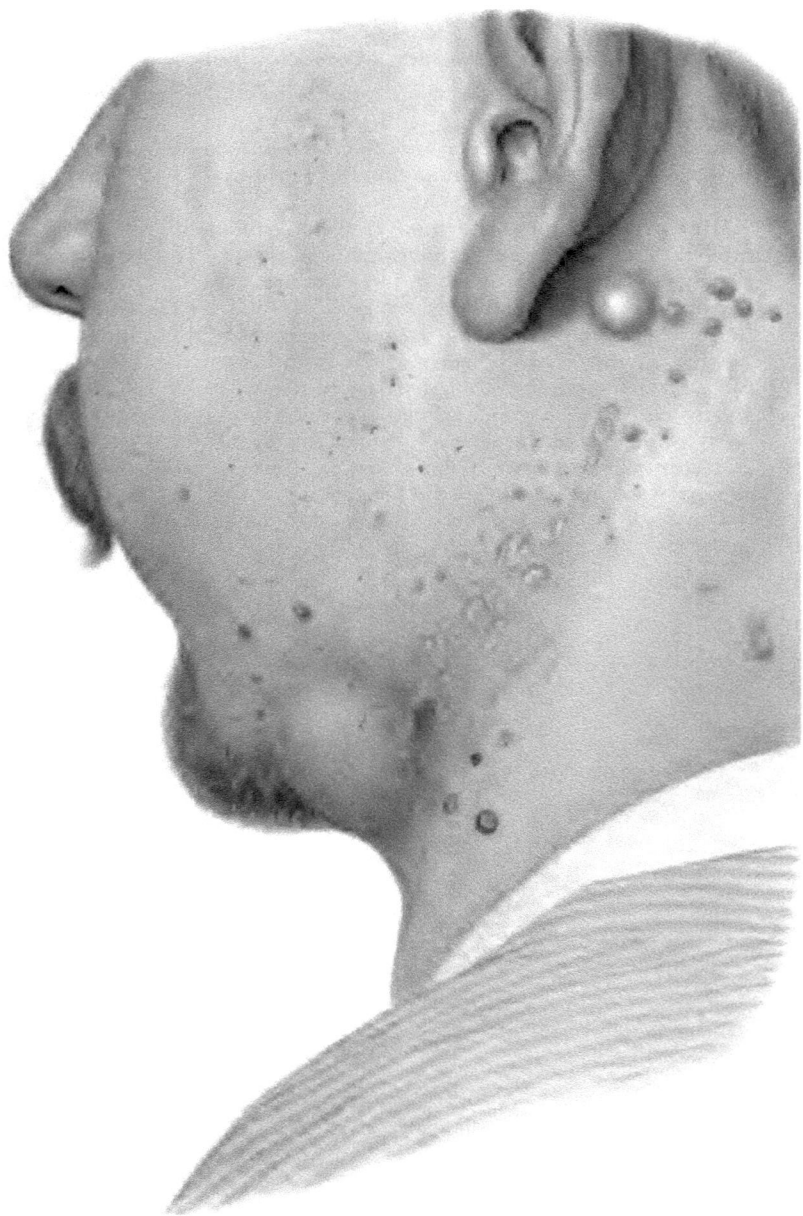

PLATE 4.

Morbilli (Papular Form).

F. F., aged 19 years, a domestic, was under observation from May 5 to 12, 1897. Patient was taken ill three days previously, with sore throat and repeated sweats; for the last day running from the eyes and an eruption on the trunk.

Status Præsens.—The patient is medium-sized, strongly built, and well nourished. The face is thickly beset with pin-point-sized reddish papules with a minute dark-red areola. On the breast and neck the eruption is similar, except that the papules are smaller and flatter and the areola less marked. While plentiful upon the breast, the eruption is wanting upon the back toward the waist. The eruption is present, but less abundantly, upon the abdomen; more profusely on the thighs and the inner sides of the knees. The lower part of the legs is entirely free. The upper extremities show the rash, extending down to the forearms. Conjunctivæ injected. Soft palate slightly red; tonsils considerably enlarged and red in spots. Patient is without fever.

Pulse and respiration normal. No subjective symptoms. Specific gravity of urine, 1011; slightly acid; free from albumin.

The patient remained free from fever, and was discharged cured in seven days, during which period the catarrhal symptoms and the eruption gradually disappeared.

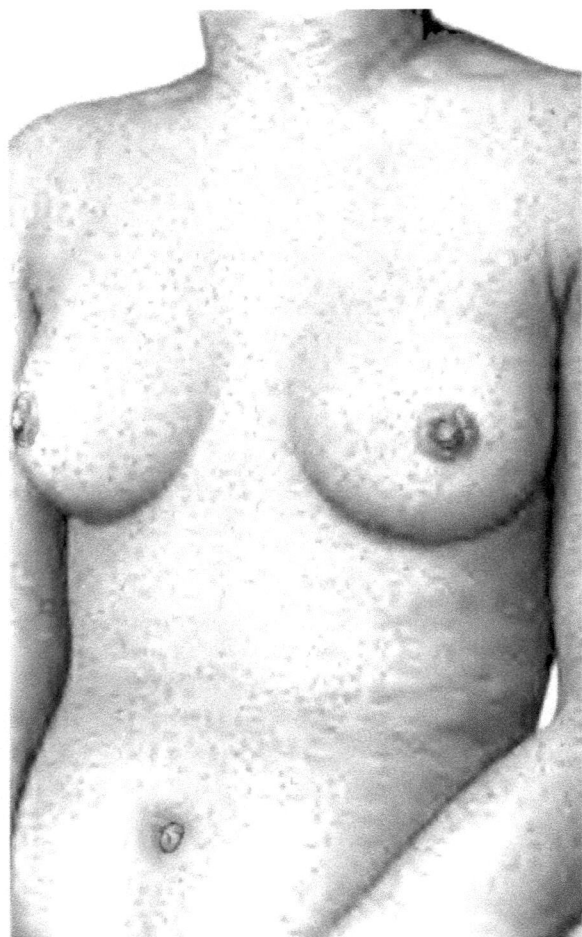

PLATE 5.

Varicella.

A. S., aged **21 years, a domestic,** admitted **on** Nov. 7, 1896, **was taken** sick three days previously with **fever,** headache, and sore throat; for the last two days an eruption had been present.

Status Præsens.—The eruption is less abundant upon the face, neck, **and buttocks than upon** the trunk and extremities. The recent efflorescences **are** miliary in size, slightly elevated above the skin-level, **and of a bright-red** color. In **their** further development they **change to rounded vesicles containing serum, and have an irregular** red border.

Nov. 12.—**The older** vesicles show seropurulent and purulent contents; **the more recent** vesicles are **still** distended with serous fluid; **all** are surrounded **with** inflammatory areola.

Nov. 16.—The vesicles have, for the most part, dried to brownish crusts.

Nov. 20.—**The** pustules have **dried up;** most of the crusts **have** fallen **off,** leaving pale-brown spots.

Nov. 26.—**Patient was** discharged cured. During the entire course **the temperature was** not materially elevated.

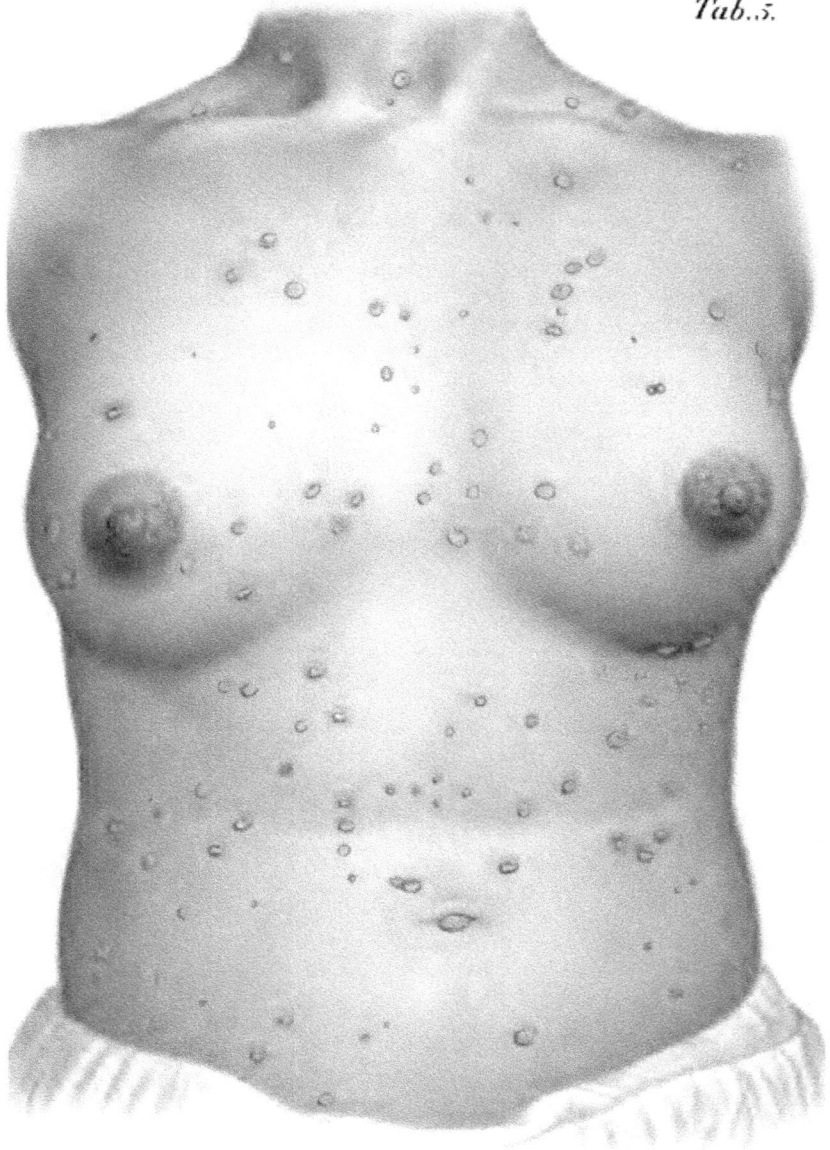

Tab. 5.

Lith. Anst. F. Reichhold. *München.*

PLATE 6.

Erythema Multiforme (Erythematous and Erythemato-papular).

G. J., aged 32 years, a waiter, was admitted Apr. 20, 1896. Two days previously, following, as the patient believed, the eating of roast pork, an eruption appeared on the face and on the hands and feet.

Status Præsens.—Patient is strongly built. An eruption consisting of bluish-red, slightly-elevated spots, with a bright-red areola, becoming pale upon pressure, is to be seen, symmetrically arranged, on the dorsal surface of both hands, the extensor aspects of both forearms, and likewise upon the lower extremities and the large toes; also the same characteristic eruption upon the forehead. These efflorescences are for the most part circular in shape, here and there several running together and forming dollar-sized areas. The palms, the soles, the mouth, and throat are free.

Following the internal administration of oil of mint and oil of eucalyptus the eruption gradually flattened without any fresh exacerbation, and disappeared with very slight desquamation. The patient was discharged cured eight days after admission.

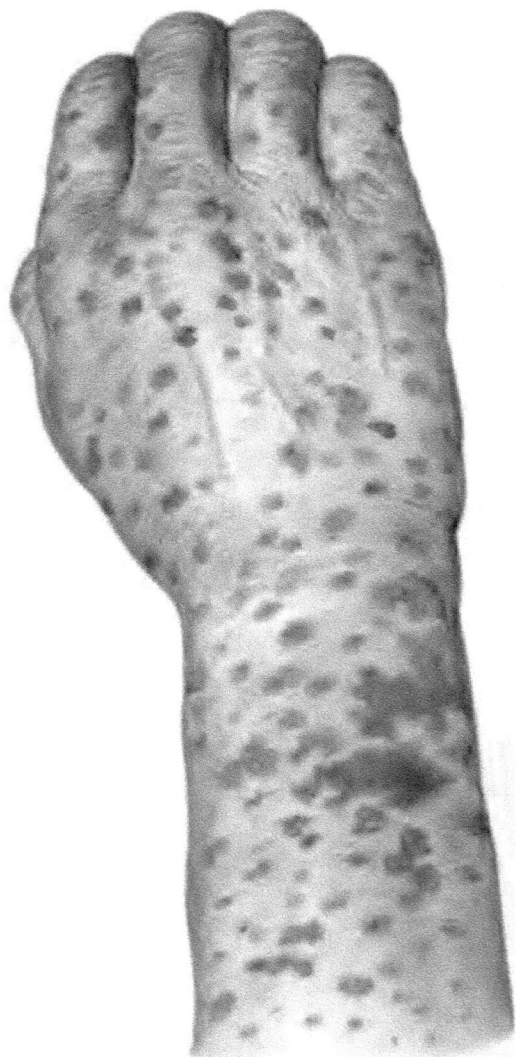

Tab. 7.

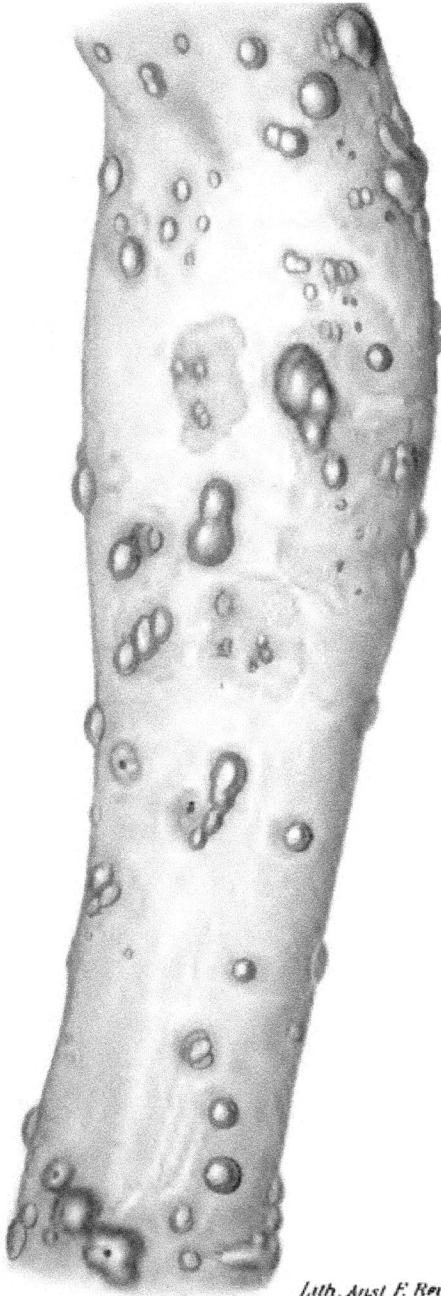

Lith. Anst. F. Reichhold, München.

Erythema Multiforme (Vesicular and Bullous).

S. A., aged 16 years, locksmith's apprentice, admitted **Mar. 18, 1897,** noticed three days previously, on awakening in the morning, an eruption consisting of small translucent **vesicles** seated upon a red base. The first lesions were observed on the axillary folds and the flexor surface of both forearms. Itching was quite marked. The individual vesicles grew larger, and new lesions appeared, in the **course of a few days, on** the trunks and extremities. During this time the patient had feelings of heat and chilliness.

Status Præsens.—Patient small, slender, with very little fat-tissue. No elevation of temperature; pulse 80, and **regular.** The urine contained traces of albumin and nucleoalbumin.

Efflorescences are to be seen on the face, especially about the chin, on the neck, profusely on the anterior thorax, on the **abdomen,** back, and upper and lower extremities. They **vary in** size from a pinhead to a silver quarter; are **pale red, rounded,** and somewhat **elevated** like wheals. They become somewhat paler on pressure, here and there leaving a yellowish tinge. **In** certain regions, as the anterior thorax, the clavicular region, and the outer side of the forearms, they have become **confluent,** forming large irregularly shaped groups and areas. In the center of many of the efflorescences **there is a blood-crust.** Near by these efflorescences, scattered over the entire surface, are countless millet-seed- to bean-sized vesicles with clear contents, and for the most part well distended. **Where** the vesicles are broken the reddish base is observed **to be covered** with dried yellowish secretion. In the neighborhood of **the left** collar-bone is an accumulation of thick hemorrhagic **crusts.** On the back are two or three blebs with hemorrhagic contents. In this region also are numerous scratch-marks. There are a few blebs on the dorsal surface of the feet. The volæ manus, the soles, lower part of both legs, and the joints are free. The mouth and throat are likewise exempt.

Mar. 19.—General condition good and no fever.

Mar. 22.—Numerous blebs filled with pus; some hemorrhagic. No new lesions.

Mar. 23.—Erythematous spots have disappeared; superficial abrasions mark the sites of burst or broken blebs.

Mar. 25.—Temperature 37.7° C. Many of the abrasions are skinning over.

Mar. 29.—Some fresh blebs on forearms and face. Temperature 38.3° C.

Mar. 31.—The abraded areas have skinned over. Highest temperature 37.8° C.

Apr. 1.—The skinned-over abrasions are still somewhat elevated. Fresh scattered and closely-crowded lentil-sized blebs with clear contents have appeared on forehead and cheeks. Temperature normal.

Apr. 4.—The blebs on forehead and face have become purulent.

Apr. 5.—Evening temperature 39.4° C.

Apr. 20.—The skinning-over process is almost complete; the epidermis on the places of former blebs is still quite red, but there is now no elevation.

Apr. 26.—Pale reddish-brown pigmentations mark the sites of the lesions.

Apr. 28.—Discharged cured.

Tab. 7 a.

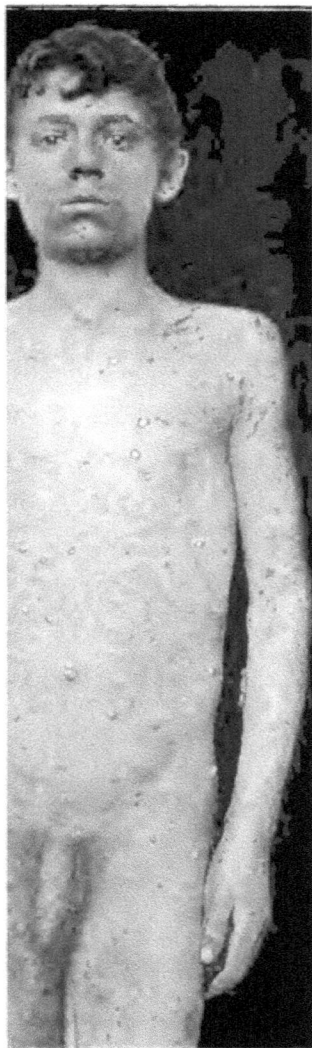

PLATE 8.

Erythema Multiforme (Papular and Nodose).

G. I., aged 11 years, admitted Apr. 20, 1896; discharged May 3, 1896. For two weeks he had noticed the appearance, without known cause, of an eruption on both arms and legs. He had previously been quite healthy.

Status Præsens.—The papules are to be seen on the extensor surfaces of the upper extremities and upon both anterior and posterior aspects of the lower extremities. The trunk is free. The eruption consists of millet-seed-sized papules, extending into the cutis and somewhat elevated above the skin-level; on their summits is, for the most part, either a minute blood-crust or -scale. In some places, and more especially in the popliteal spaces and over the patellæ, are observed dime- to shilling-sized bluish-red nodes (erythema nodosum).

Treatment.—Sodium citrate.

In the course of the disease there was slight hemorrhage into the disappearing papules, which, however, was rapidly absorbed. The nodose lesions gradually disappeared, undergoing the usual color-changes.

Tab. 8.

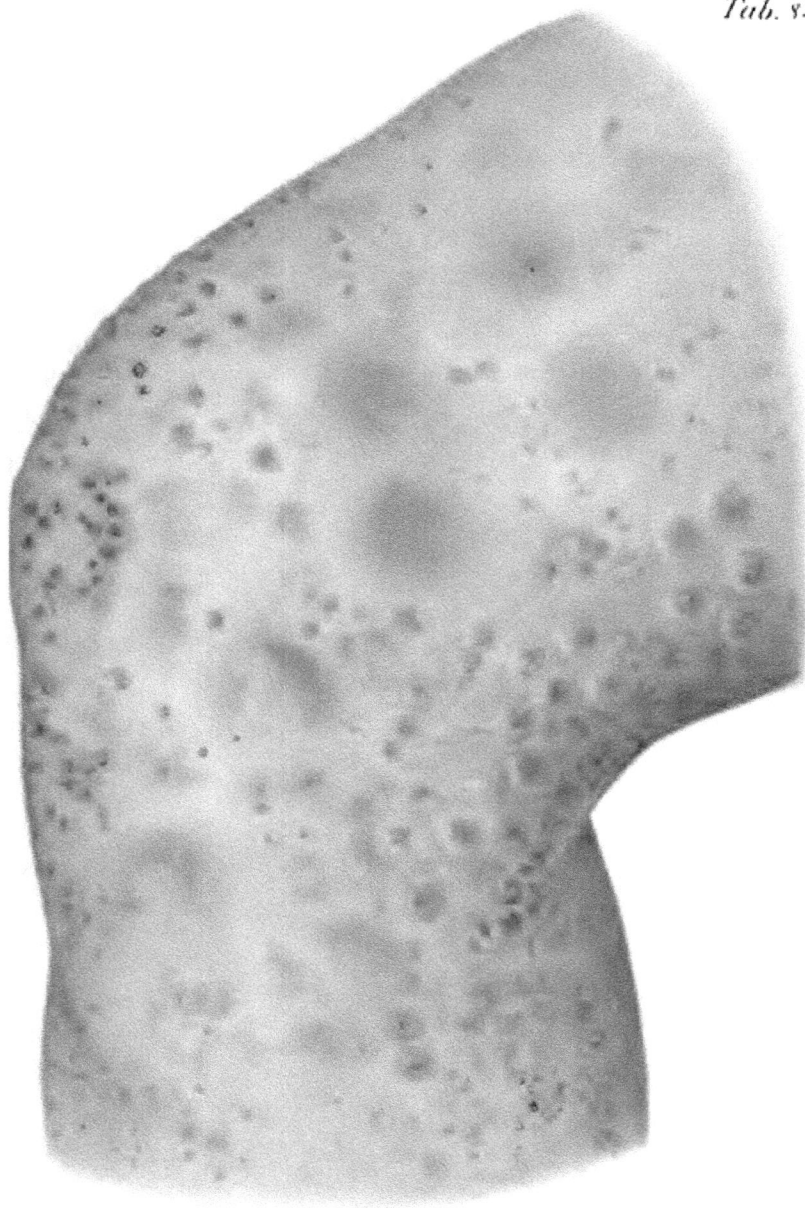

Lith. Anst. F. Reichhold, München

PLATE 9.

Purpura Hæmorrhagica.

M. M., aged 33 years, coachman, admitted Apr. 23, 1897, stated that for the last eight days he had felt exhausted and sick, and had observed spots in the skin. Similar spots he had noticed several times previously; but as they had disappeared without discomfort or medical aid he had never considered them of any moment. He sought the hospital this time owing to the feeling of general weakness and depression.

Status Præsens.—Patient is large, well built, but pale. The gums are livid and furrowed, and bleed easily; conjunctivæ jaundice-colored. The heart-tones are somewhat dull; pulse 84, and soft; spleen not enlarged. There is neither sugar nor albumin to be found in the urine. Over both ankles and on the dorsal aspect of both hands there was slight edema. On the lower extremities, about the hair-follicles, are pinhead- to lentil-sized recent and old hemorrhages, here and there showing a tendency to be closely set together and in rows. In addition to these lesions are to be noticed rounded and more or less diffused violaceous spots on the lower legs, in the central part of which the follicular hemorrhages are more crowded. In these latter places the skin has a succulent feel. Scattered hemorrhages are also noticed on the trunk.

Apr. 30.—Urine shows considerable urobilin. In sediment, hyaline cylinders and a few blood-corpuscles.

May 1.—On the inner side of the upper part of both thighs, especially the right one, fresh follicular hemorrhages of a reddish-brown color, and bluish, livid spots have appeared; the latter are so extensive as to become confluent.

May 5.—The brownish-red hemorrhages and the livid spots begin to change to a yellowish tinge. The patient suffers pain in the legs.

June 1.—The calf-muscles feel hard, and the patient when attempting to walk has considerable pain in these parts. The gums are still swollen, the inner side showing numerous minute hemorrhages.

June 20.—There is considerable pain in knees and hips. Evening temperature rises to 39.5° C.

July 1.—The hemorrhagic spots on trunk and thighs have, for the most part, been absorbed. The calves are softer to the feel.

Aug. 19.—The diarrhea, which had existed for some time, alternates with constipation. Urine-examination shows considerable indican and skatoxyl. There is a somewhat painful swelling, soft in character, about the knees; the overlying skin unchanged.

From now on no new lesions in the skin were observed. The patient still suffered, however, from effusion about the knee-joints, marked debility, and inability to walk any distance.

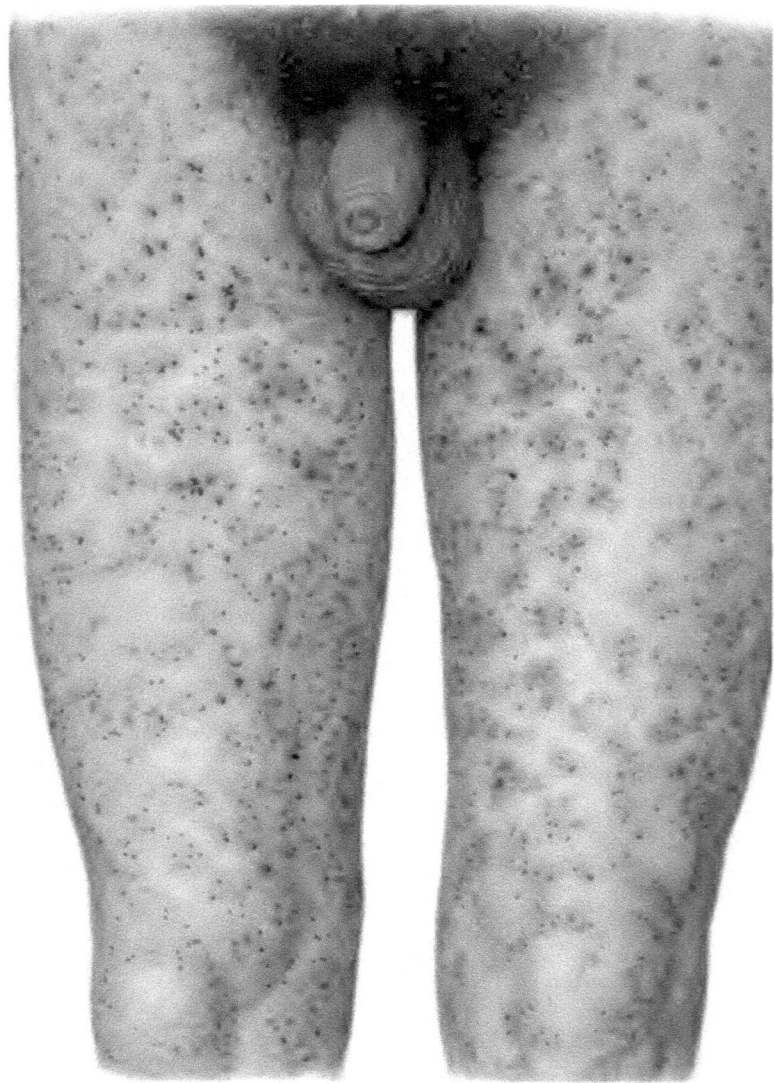

PLATE 10.

Purpura Hæmorrhagica.

S. M., aged 16 years, working-girl.

Status Præsens.—Patient small and spare; pallor of skin and mucous membranes, and slight enlargement of the heart toward the right. No pulse-irregularity. Menstruation is not yet established. Hemoglobin, Fleischl, 55 per cent.

For four days, beginning on 17th, the patient had noticed hemorrhages in the lower extremities. She worked in a laundry, standing during the whole day.

On the lower extremities, from the middle of the thighs and extending down over the entire lower legs, are lentil- to pea-sized, scattered and confluent cutaneous hemorrhages, some of which already show a change to a brownish tint.

24th.—Eight days after the beginning of the outbreak the general health seems good; the efflorescences yellowish and some becoming skin-color.

Tab. 10.

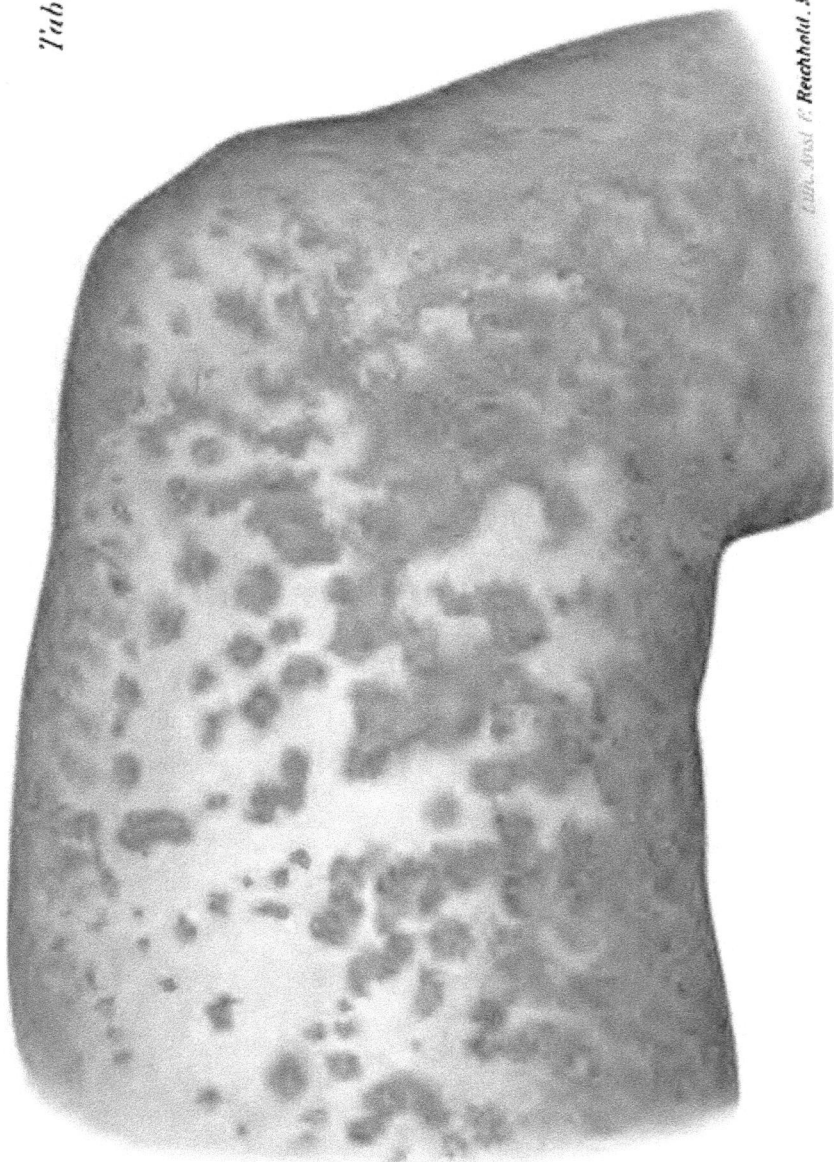

Lith. Anst. C. Reichhold, Munch.

PLATE 11.

Purpura Rheumatica (Fulminans).

J. M., aged 38 years, clockmaker, admitted Sept. 12, 1897.

History.—Nine years previously patient had a pleuritis. For the past four years has had attacks of pain in the large toes, lasting four to six weeks. Has been addicted to drink. His present disease began on Aug. 28 of this year, with stinging in the heels and the appearance of small red macules on the lower extremities. The pain and the spots disappeared in the course of several days. There soon followed pain in the knees; later in the elbows, hands, and finger-joints, accompanied with swelling of the painful parts. At this time the patient noticed the appearance of dark-brown macules on both forearms; these spots increased rapidly in size, exhibited superficial vesicle-formation, and gave rise to marked tenderness and pain. Patient was debilitated, feverish, and without appetite.

Status Præsens.—The man is large and strongly built, with well-developed panniculus adiposus. Lungs, heart, and abdominal organs apparently normal.

The face is decidedly reddened; the left cheek is somewhat infiltrated; on the latter some lentil- to bean-sized spots, violet to blackish in color, without elevation or tenderness on pressure. On the left ala nasi is a reddish-brown hemorrhage. The entire outer border of the right ear is hemorrhagic, of a blue-black color, and very painful. The mucous membrane of the mouth and throat is normal. The right upper extremity is swollen and both at shoulder and elbow held in flexed position. On the outer side of the arm is a dime- to quarter-dollar-sized patch of dark-violet skin; the overlying epidermis is elevated as in a blister, and the whole area is surrounded by a red areola. Besides this, several painful, partly pale-red and partly dark-red spots are to be seen near by. On the inner side of the arm is a palm-sized dark-violet area similar to the large spot just described. Toward the axilla is a spot which already has begun to change to a yellowish color.

The joints of the left arm and forearm are also swollen and tender, and present similar, but smaller spots. Both knee-joints are swollen and tender and the skin reddened. Over the ankle-joint are several spots, which had already become

brownish. The other joints of the leg and foot are not swollen. The inguinal glands are unaffected.

Urinary examination disclosed specific gravity 1020, some albumin, but nothing else abnormal. An examination of the blood showed a slight leukocytosis.

Treatment.—Sodium salicylate 4 grams (60 grains) pro die.

Sept. 14.—Numerous elevated red spots, with pale-red areola and markedly itchy, are to be seen on the outer side of the right and anterior aspect of the left thigh. Temperature up to 38.1° C.

Sept. 17.—Numerous lentil- to pea-sized, elevated, hemorrhagic, itchy papules have appeared on the thorax. Temperature 39.2° C.

Sept. 20.—Hemorrhage has taken place into the left earmuscle; the left foot is edematous. Passing large quantity of urine.

Sept. 22.—The joints are again much swollen. Fresh outbreak of hemorrhages on thorax, abdomen, and thighs. Temperature 38.7° C.

Sept. 27.—The flexor muscles of the right forearm, close to the elbow-joint, markedly inflated and painful. Fresh hemorrhages in the left loin. On the hemorrhagic patches on the upper extremities superficial ulcers have appeared.

Oct. 3.—The hemorrhagic spots are being slowly absorbed. The swelling and tenderness of the joints are fast disappearing. Temperature normal. The ulcers on the arm are granulating. Zinc-oxid salve ordered.

Oct. 19.—Albumin no longer in urine. The ulcers on the left arm completely healed; on the right are still two small ulcerations covered with abundant granulations.

Oct. 28.—The affected areas are seen to be slightly pigmented. Patient was discharged cured.

PLATE 12.

Herpes Zoster (Sacrolumbalis, Hæmorrhagicus et Gangrænosus).

S. W., aged 66 years, female, admitted Feb. 29, 1896, stated that for two months she had suffered from bronchial catarrh. For the past fourteen days she had pain in the back and thigh. A few days ago an eruption appeared on the buttock of the right side, accompanied with burning pain, and rapidly became more and more extensive.

Status Præsens.—On the right buttock several groups of herpetic efflorescences, with hemorrhagic contents and with hemorrhagic areola, are to be seen. Some groups have become gangrenous and changed into shallow ulcers with hemorrhagic base. Over the region of the sacrum are recent vesicles with serous, and partly milky, contents. Six days after admission, and the twentieth day after the beginning of pain, the lesions, began to dry up and the tenderness and pain were much less marked.

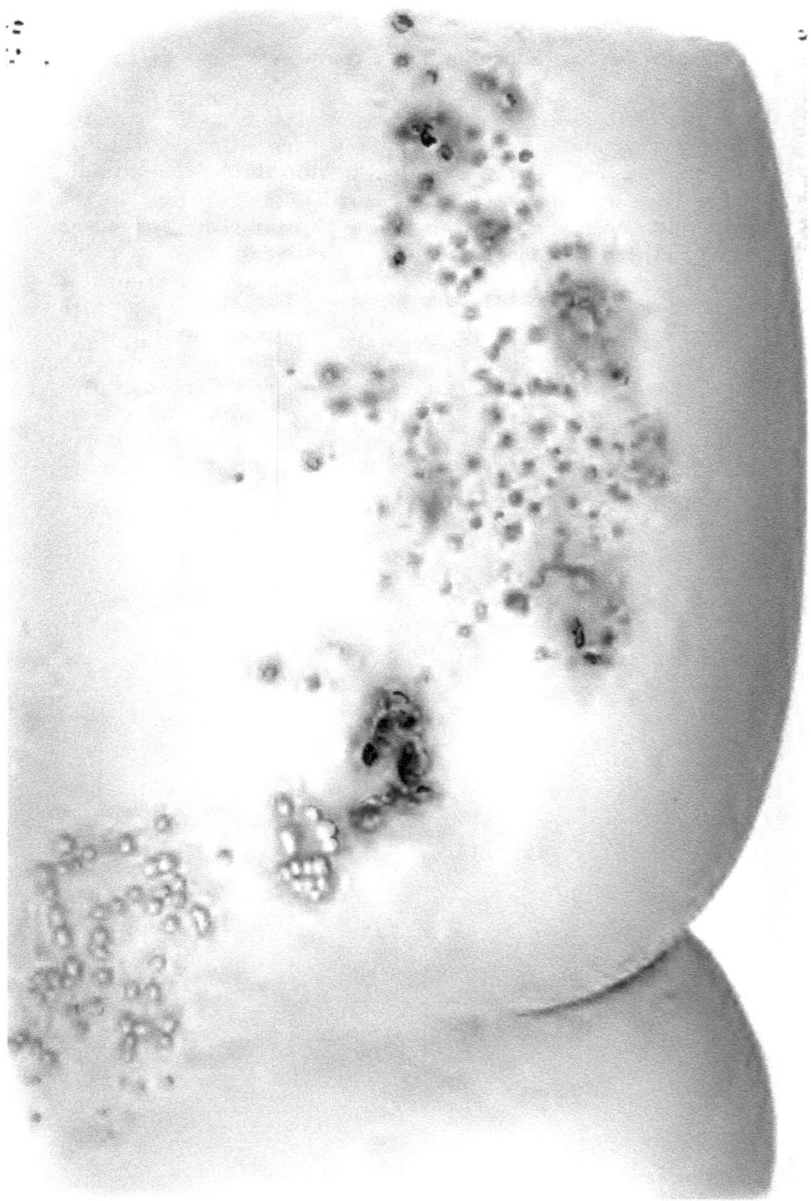

PLATE 13.

Herpes Zoster (Supraorbital and Palpebral).

L. S., aged 16 years, mechanic, came of a good, sound family and was himself always healthy.

In Mar., 1896, the patient had a similar eruption on the same region for a period of eight days; since then he has had no sickness; never had headache or other nervous symptoms. Five days ago the patient felt unwell, had a chill, and toward afternoon felt obliged to lie down. On the following day the left upper eyelid was red and swollen, and on the next morning he noticed some vesicles upon the nose and eyelid toward the inner angle. Yesterday, four days after the first symptoms presented themselves, vesicles also appeared on the eyebrow; and this morning early the two recent groups which cover the outer side of the lid. There is a feeling of distention and burning in the affected lid.

Status Præsens.—The eruption is to be seen on and around the inner side of the lid and on the nose, and consists of vesicles with reddish base and areola, and beginning to dry. The entire upper lid is swollen and edematous, inflamed, and red. On the brow is to be seen a group of greenish-yellow vesicles, tending to become confluent. In the middle of the lid and also toward the outer side are two groups of recent, yellowish-white vesicles, partly confluent. Toward the edge of the lid and upon the border are pinhead-sized scattered vesicles.

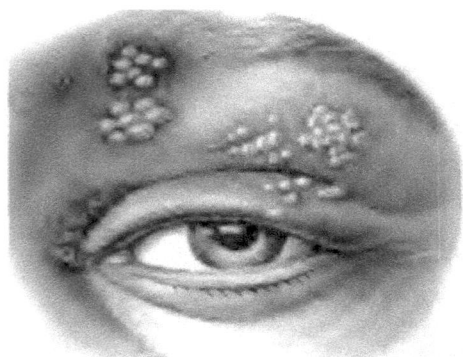

PLATE 14.

Dermatitis (Cantharides).

S. M., aged 35 years, drug-clerk, healthy, had from morning till mid-afternoon worked with cantharides, sifting it, and perspired freely during this time. Toward evening he felt burning sensations; blebs appeared, which enlarged considerably during the night.

Status Præsens.—On the forearms and on the neck are irregular, well-distended blebs with serous contents, and having reddish areola. The rest of the body is free.

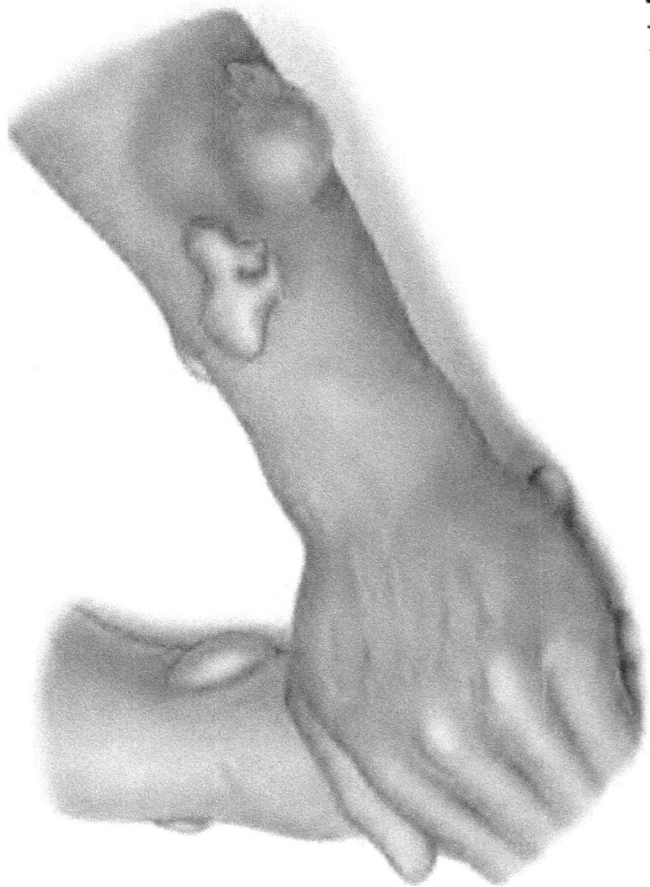

PLATE 15.

Psoriasis (Punctata et Guttata).

G. J., aged 17 years, locksmith, was admitted Mar. 23, 1886.

The patient had his first attack of more or less generalized psoriasis two years previously, which, with the exception of a few spots on the knees, had entirely disappeared after the use of salves. The present eruption was noticed fourteen days before admission, first on his arms.

Status **Præsens.**—The patient is strongly built, well nourished, and is apparently in good general health. On the trunk, extremities, and face are numerous psoriatic efflorescences, varying in size from a mere point to a lentil, having the characteristic scaliness. On the extensor surfaces of the knees are larger lesions, apparently of longer duration.

The painting was made two weeks after the beginning of the present outbreak. A week later the lesions were more numerous, with a tendency to form confluent patches, and covered with silvery scales. The borders of the patches were bright red.

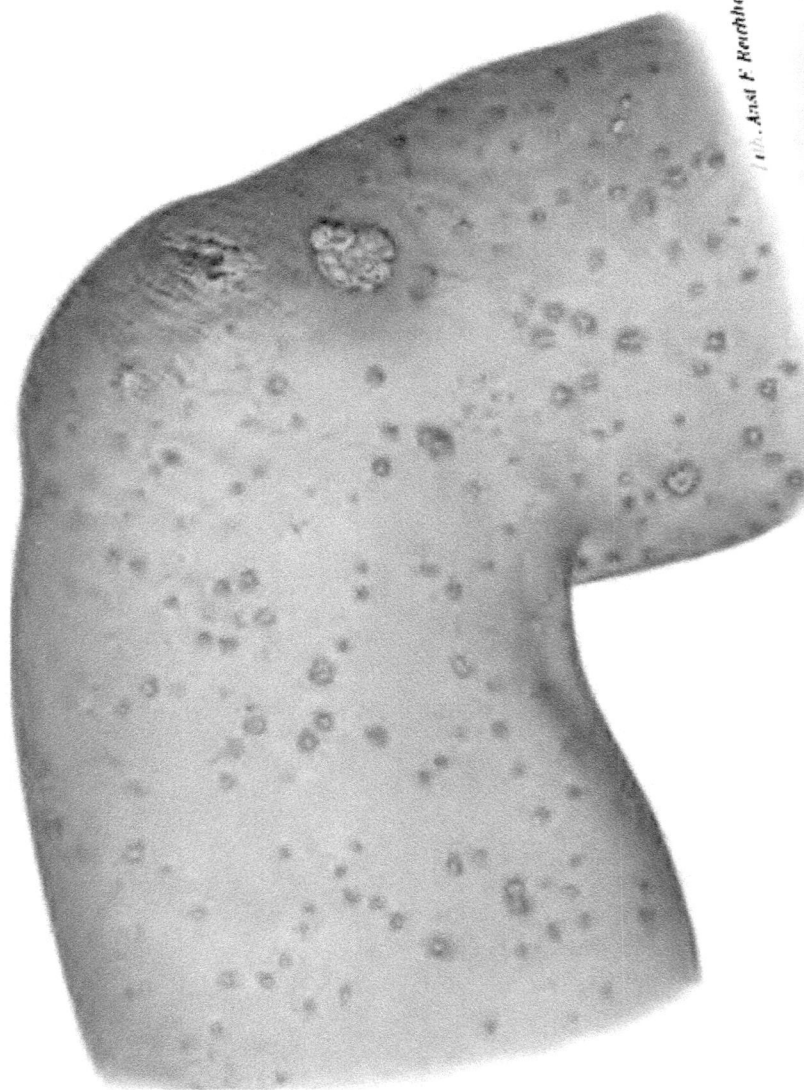

Lith. Anst. F. Reichhold Munchen

PLATE 16.

Psoriasis (Diffusa).

H. C., aged 41 years, shoemaker's helper, admitted Aug. 22, 1896, stated that four years previously he had been treated for the same disease. For the past five months he noticed a reappearance of the eruption. He had himself made applications of petroleum, but with no result, and had then sought the hospital.

Status Præsens.—The entire surface of the patient is covered with psoriatic patches. In some places they have become confluent, forming large red infiltrated areas, covered with scales. This is more especially the case on both lower legs, on the outer sides of both thighs, in the lumbar region, and on the extensor surfaces of both forearms. On the scalp the eruption is extensive and confluent. On the trunk, on the chin, and on the forehead are scattered lesions, pinhead- to pea-sized. All patches are moderately elevated with narrow red border, and covered with fine white scales.

Treatment.—Chrysarobin salve. After ten days' use, owing to a conjunctivitis, this was temporarily discontinued. After the conjunctival inflammation was relieved treatment with the salve was again begun, and the case finally cured.

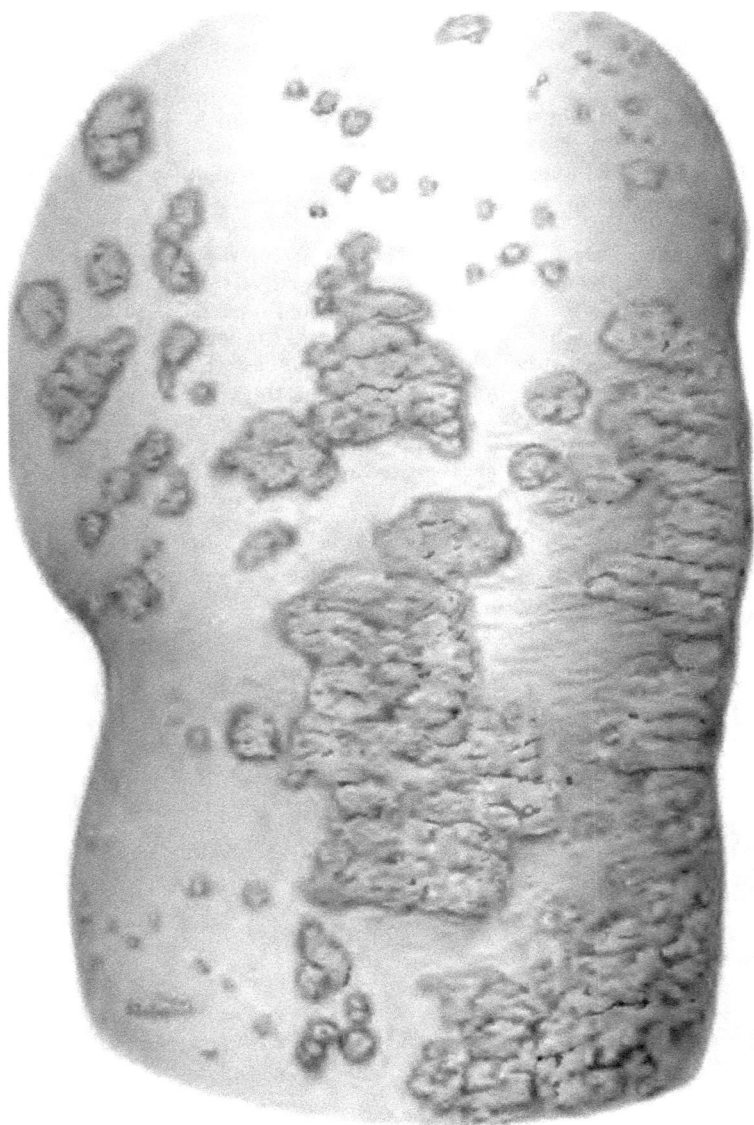

Psoriasis Nummularis [Eczema Seborrhoicum ?--Ed.].

S. J., aged 64 years, vine-grower, was admitted Feb. 5, 1896. He stated as follows : That he was always healthy ; in 1868, without any apparent cause he rapidly lost his hair. Is a moderate drinker. His present disease was first noticed about a half year before admission, and first on the trunk and hands. After moderate itching some blisters appeared, which dried to crusts. For a long time he had been in the habit of removing these crusts with oil, but they always reappeared. Later the eruption appeared on scalp and face.

Status Præsens.—Patient is strongly built, but not well nourished. The scalp is covered with brownish-white crusts, which when loosened can be made to come off as an ill-defined cast. The hairless underlying skin is thin and hyperemic.

On the face, about the eyelids, are eczematous, weeping patches. Chronic conjunctival catarrh, moderate ectropium of the lower lids, and increased tear-flow are noticeable. On the breast and upper belly-region and on the extremities the eruption is extensive, consisting of scattered half-dollar-sized, palm-sized, and larger areas. The scattered spots show in the central portion considerable scaliness, more or less heaped up, and have a hyperemic border. Vesicular formation cannot be seen on any part.

The paintings show two of the more recent patches on the breast (Plate 17), and on the scalp (Plate 18) after partly freeing it from crusts.

[By many these plates and description would be considered to belong to cases partaking of the nature of both eczema and psoriasis (psoriatic eczema), and by others as eczema seborrhoicum—Ed.]

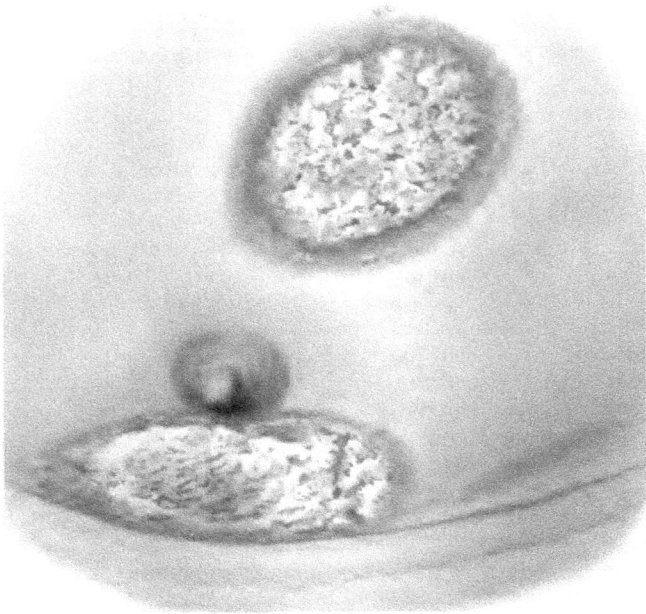

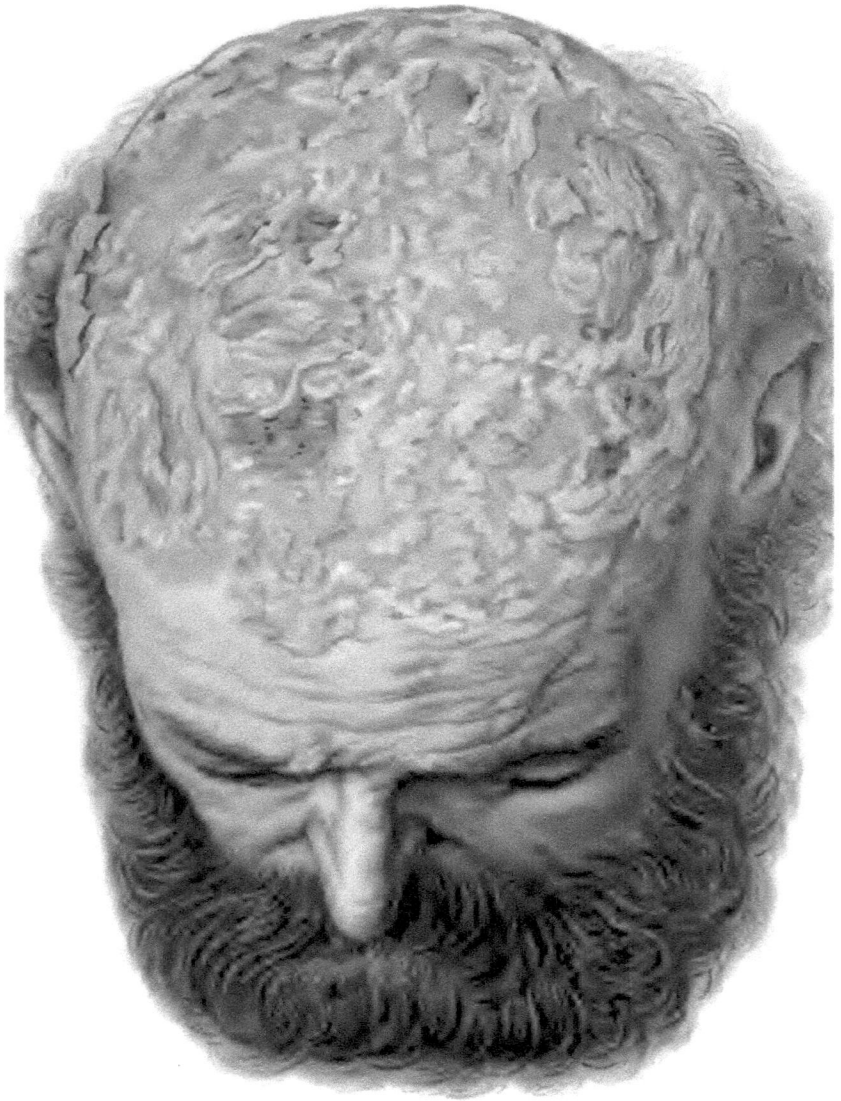

Psoriasis (Circinate, Annular, Gyrate).

H. F., aged 21 years, laborer, admitted July 6, 1896, stated that for a period of eight years scaly papules had been observed. Two years ago he was treated elsewhere with tar-tincture and drops (arsenic?), and later, in winter, for three months with pyrogallol, chrysarobin, and pills (arsenic?), but without result.

Status Præsens.—Patient is medium built, with fair bone-structure and muscular development. Internal organs normal. On the buttocks and on the upper extremities are scaly papules, pinhead to lentil in size and with a narrow red border; likewise larger efflorescences, rounded and with infiltrated bases. On the abdomen and upper third of the lower extremities are patches consisting of an infiltrated, red, scaleless center, surrounded by an annular border covered with glistening white scales.

Course and Treatment.—Thyroidin was prescribed in capsules, each containing 0.50 gm. (7½ grains), beginning with one daily, and increasing one every three days. The pulse was not materially affected, and the body-weight varied but several pounds (between 52 and 57 kg.). The skin-condition gradually improved, so that on Aug. 19 the following status was noted:

On the arms the psoriasis-spots are pale; infiltration and scale-formation have disappeared. The circinate and gyrate patches on the breast, abdomen, and back exhibit less redness, being now pale red or brownish, with retrogressive infiltration and scale-formation. Pulse 100, and regular.

Early Sept.—On buttocks are still to be seen some slightly elevated areas, and some patches which are still somewhat red.

Sept. 19.—The patient was discharged cured.

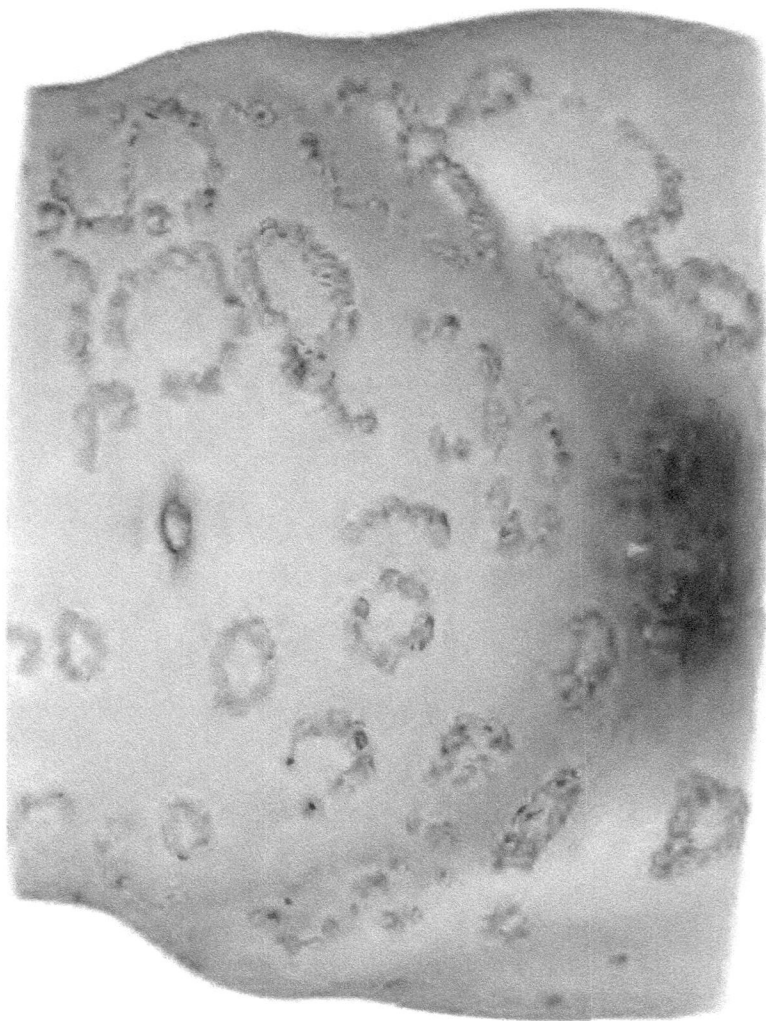

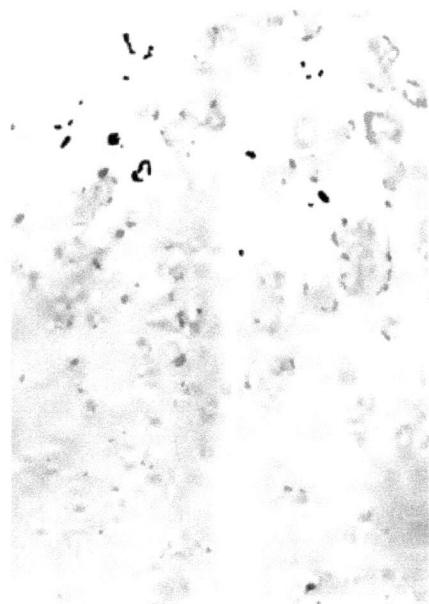

PLATE 20.

Psoriasis (Gyrate, Annular).

C. F., aged 21 years, laborer, admitted Mar. 18, 1897, stated that he had psoriasis for the first time in 1893, at which period the patches appeared on the extensor surfaces of the elbows and knees. Under treatment with pyrogallol and chrysarobin salves he was much benefited. A year ago he noticed a change in the diseased areas—spontaneous disappearance of the central portions and an extension and confluence of the borders.

Status Praesens.—Patient is of graceful build; moderately nourished. Internal organs normal. On the legs, arms, and trunk, in addition to scattered pinhead-, pea-, and coin-sized lesions, are to be seen large serpentine or irregularly circinate plaques, the peripheral portions being made up of hyperemic, elevated, scaly borders, sharply defined, and enclosing areas of brownish pigmented skin. Here and there within these boundaries are to be observed lentil- to pea-sized scaly spots. The scalp is reddened and covered with thick scales. Body-weight (Mar. 17), 54.5 kg.

Treatment.—Iodothyrin.

Apr. 6.—Weight, 52.1 kg.

Apr. 16.—Patient was, upon request, discharged, some improvement having taken place.

Tab. 20

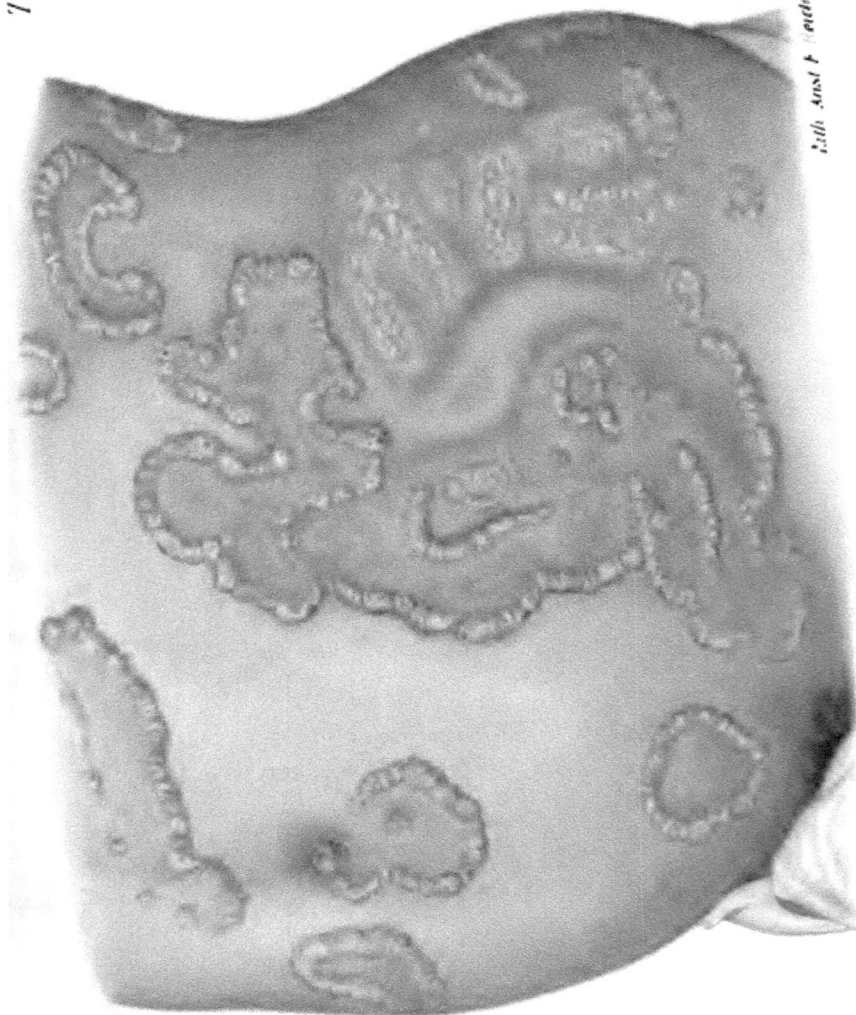

Lith. Anst v. Rodthold Mü.

Psoriasis. Cornua Cutanea (with Degenerative (from Uric-Acid Diathesis) Changes in Right Hand and Left Foot.

H. J., aged 58 years, an innkeeper, was admitted May 5, 1897. The patient stated that his father had been a sufferer from gout, and that he himself, when in his thirty-third year, was ill. His illness began with symptoms of general weakness, which increased, and was accompanied with swelling of the joints of the feet. This condition lasted some months. In 1883, when about forty-four, he again became sick, and was obliged to keep in bed; there were swelling and pain in all joints, especially those of the lower extremities, and in the loins. Four years later he had a similar attack. In 1891, there developed a tumor or swelling on the head, which was removed by operation. In 1889, scaly papules appeared on the right shoulder, since which time similar lesions had made their appearance on the trunk and extremities. The hands were free up to three months before admission, when the eruption appeared on these parts; there was pain in the right hand. Lately the patient had lost considerable flesh. Appetite was good. The bowels were sluggish, sometimes five days elapsing between the stools.

Examination of the urine passed in twenty-four hours showed a marked increase in the uric acid and considerable uric-acid sediment. Urine was much less actively solvent for the uric acid than normally.

Status Præsens.—Patient is large, pale, very much emaciated, and of delicate bony structure. Pulse 63; rounded and well filled. Temperature normal. Arteries hard. Lungs emphysematous, and disclosing many râles and much whistling. Heart-sounds apparently normal. Liver and spleen could not be made out.

The skin in general is dry and easily lifted in folds, the subcutaneous fat having disappeared. On both forearms the skin is parchment-like and in wrinkles and folds; on the thighs the folds are thicker and more marked. Over the general surface, with the exception of the face, neck, breast, and the back down to the sacrum, are to be seen innumerable lentil- to palm-sized scaly patches (psoriasis guttata et nummularis). In certain places, as on the buttocks and lower legs, the eruption has

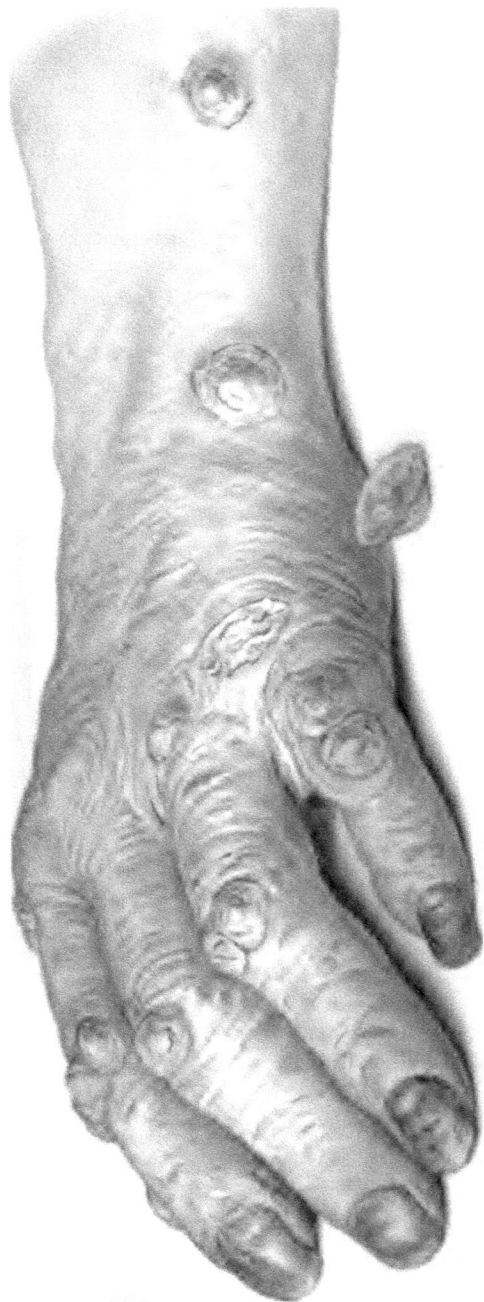

become confluent and formed festoons. Over the olecranon, left arm, is a chestnut-sized, rounded, closely-adherent, heaped-up, shell-like scale, surrounded by a red infiltrated border. Similar lesions are to be seen, with smaller crust-formation, heaped up and rounded in form, on the forearms, hands, and lower extremities. Upon lifting the shell-like accumulation from these lesions, the papillæ are disclosed, the surface bleeding easily. On the extensor surface of the right elbow the eruption is of the usual character. On the extensor surfaces of both knees are yellowish crusts seated upon grater-like, raw-looking skin.

On the dorsal aspect of the second joints of the fingers of the left hand are also heaped-up, oyster-shell-like scaly crust-formations; in consequence of which the fingers are held in a bent position and cannot be extended—the stiffness of the joints of this part is, however, partly responsible. The nails of these fingers are thickened, of dirty gray color, fissured lengthwise, and lifted up from the matrix by a horny accumulation beneath. The right hand (Plate 21 a) and fingers, especially on the dorsal aspect, are considerably swollen, reddened, and infiltrated. The palms are the seat of yellowish, tough, hard, horny scales. The nails of the right hand jut out, talon-like, over the finger-ends, and rest upon a horny, hypertrophic nail-bed, although less so than with the nails of the other hand.

The large joints of both big toes are pushed forward, and bent, valgus-like, and covered with horny masses. Similar horny accumulations are to be observed on the soles. The toe-nails are irregular; in part wanting, in part showing horny masses.

Course and Treatment.—In the further course of the disease the patient complained of pains in the hand-joints and of a troublesome cough. Treatment consisted of Carlsbad cure, milk-diet, and baths. Under this treatment most of the crusts and scales had in four weeks' time fallen off.

June 14.—The horn-like psoriatic accumulations on the elbows, lower legs, and around the ankle have been cast off; the borders are still red and slightly scaly. The joint-affection has considerably retrogressed; the nails have hardened, are thickened, bent, cracked, and exfoliating. The patient's general appearance is materially improved, so that in this improved condition, at his own request, after a period of six weeks' treatment, he was discharged.

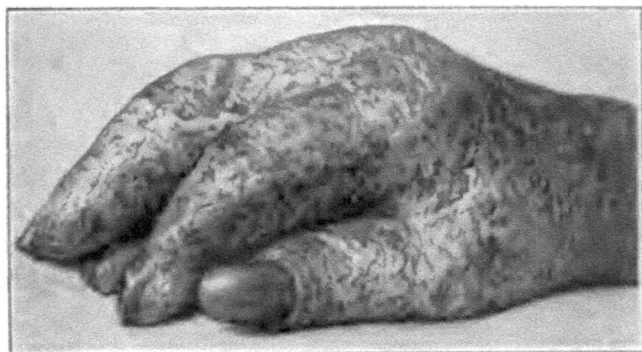

Tab. 21 a.

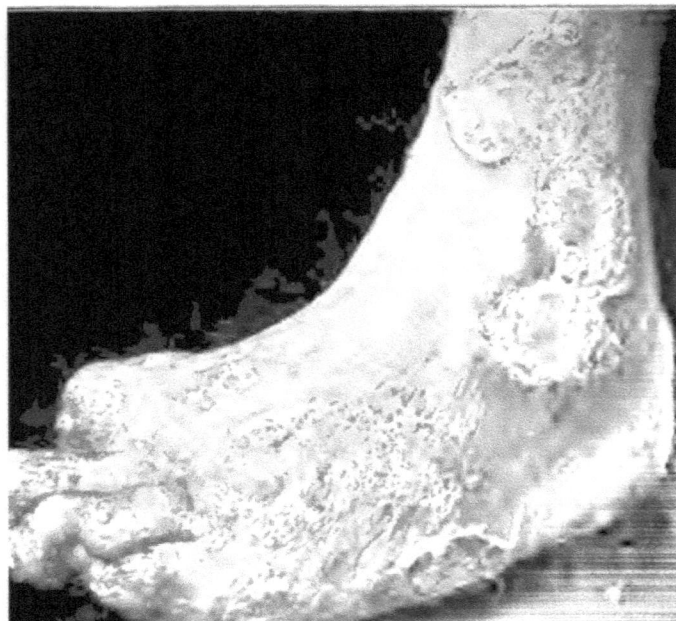

Tab. 21 b.

Lichen Ruber Planus.

U. S., aged 41 years, female.

The eruption is somewhat widespread. The face is free. On the upper extremities the flexor surfaces are more especially involved, the lesions on the extensor surface being scanty and scattered. On the lower extremities the anterior surface of the inner side of the thighs and the flexor surface of the lower leg are most affected. On the back and breast and the inner side of the thighs the individual lesions making up the patches and areas are less recognizable, owing to their being confluent, the normal skin between appearing as irregular, narrow spaces. On these parts the diseased areas are of an even copper-red with a brownish tone, covered here and there with small adherent white scales. As the sound skin is approached the individual character of the lesions making up the confluent areas is readily recognized. Such lesions are red, follicular, millet-seed-sized, somewhat firm papules, becoming paler upon pressure. On the top of each is a minute scale of epidermal exfoliation. In some places the patches are somewhat masked by the effects of scratching and covered with hemorrhagic crusts, and the eruption rendered somewhat dull and less shining in character. The mouth is entirely free.

Treatment consisted in the administration of Asiatic pills, and externally salicylic acid and resorcin salves.

Tab.22

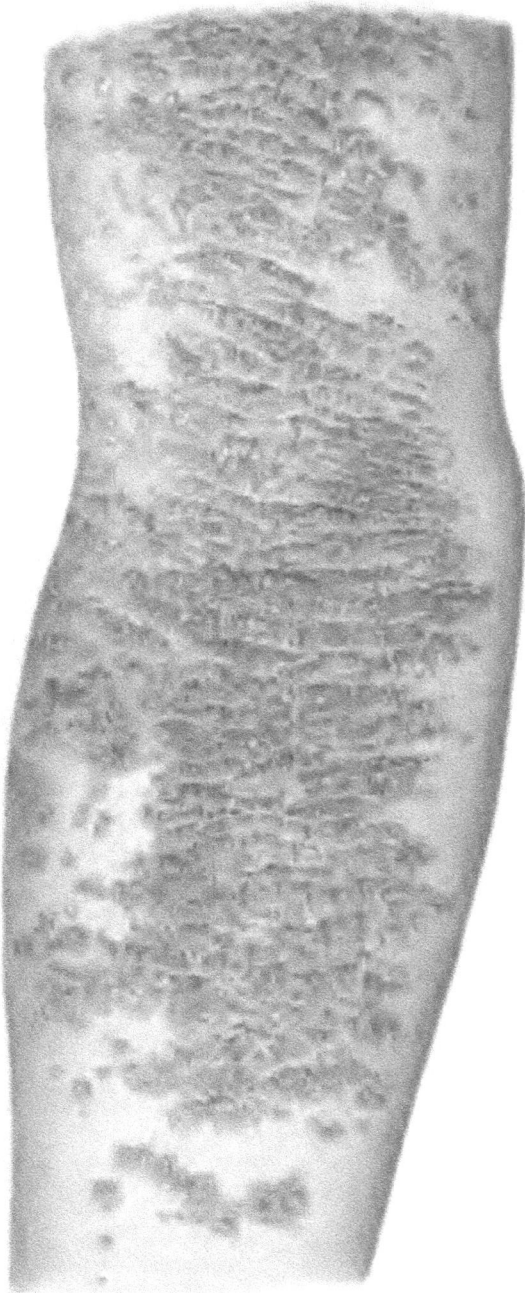

Lith. Anst. F. Reichhold München

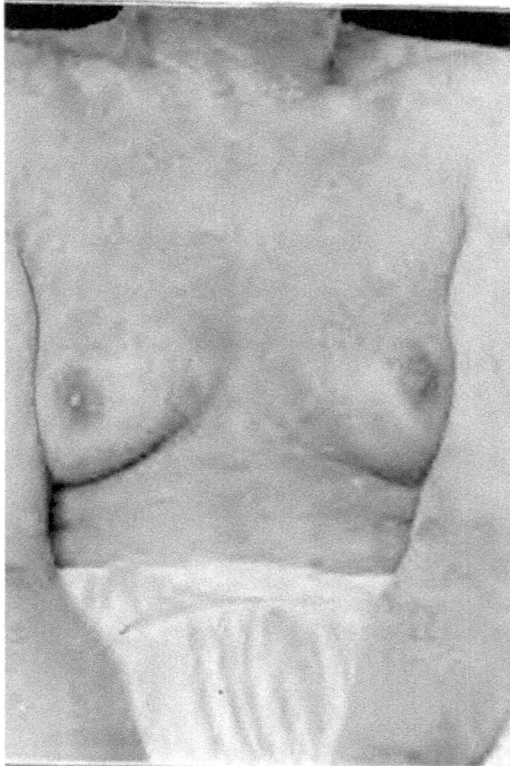

PLATES 23 and 23 a.

Eczema Artificiale Vesiculosum [Dermatitis—Ed.].

Ch. K., admitted Jan. 1, 1896. The patient was, when admitted, the subject of scabies. On the 16th and 17th he rubbed in naphthol-soft-soap. On the 19th he fell sick with fever; temperature, 38.2° C.; evening, 39.1°, C. The skin became eczematous, and of chiefly vesicular character. On the 29th the morning temperature was 38° C., and the evening 39° C. The vesicular lesions of the eczema persisted.

The urine-examination disclosed a large quantity of albumin. On the 21st the temperature fell to 37.1° C. and the vesicles had for the most part dried. On the outer aspects of the thighs, where the eruption is less pronounced, are irregularly-scattered papules, which have partly dried into thin scales or crusts, and partly show a cracked epidermic covering. The anterior aspect of the leg is covered with yellow vesicles, with light-red areola.

The size of the vesicles varied from that of a pinhead to a lentil. On some places they have become confluent and form irregular clusters, in some of the lesions and groups the epidermal covering being lifted up by the abundant pus. On the inner thighs the eruption has dried into yellowish crusts of shining aspect and is irregularly divided into areas with whitish lines (cracks).

Jan. 22.—All the pustules have dried up and the inflammatory symptoms disappeared. Patient feels much better and is more comfortable.

The painting was made from the middle portion of the thigh, from both the inner anterior and external aspects. The dermatitis evidently resulted from the naphthol. This being absorbed, irritated the kidneys, so that in the beginning a large quantity of albumin and naphthol could be demonstrated.

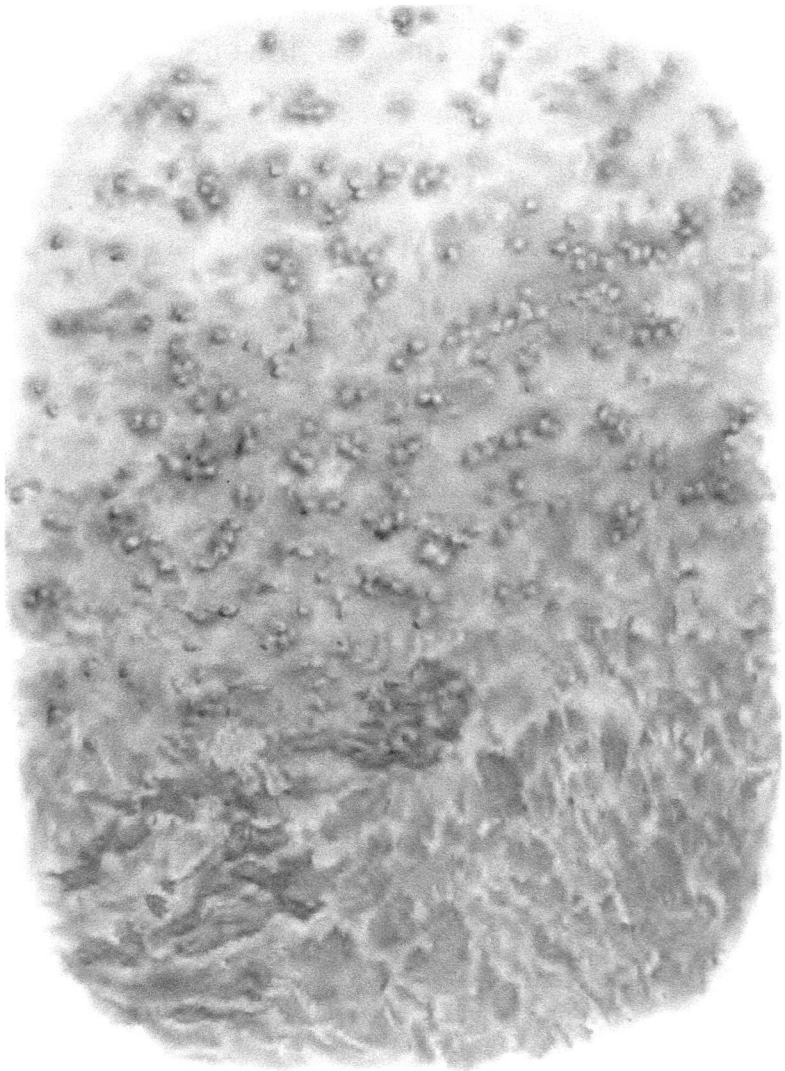

Tab. 23 a.

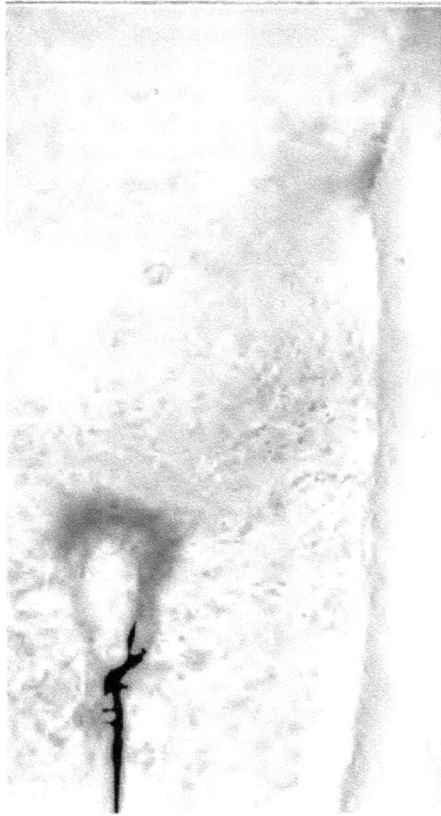

PLATE 24.

Eczema Artificiale Acutum [Dermatitis—Ed.].

Sch. J., aged 47 years, worker in the arsenal, was admitted Aug. 6, 1896. Patient was burnt on Aug. 5 by a hot piece of iron falling on him, producing burns of the neck, hands, and thorax. In the beginning he was bandaged with iodoform-gauze, and then treated with lime-water and oil.

Status Præsens.—On the neck and right forearm down to the wrist are burns of the first and second degrees. On the left side of the chest is a diffused redness. Temperature and pulse normal. No constitutional symptoms. Boric-acid salve was used. For some inexplicable reason, at the suggestion of a hospital-helper, he rubbed some naphthol salve on the mucous membrane of the lips, and immediately afterward an erythema spread over the trunk. At the same time the whole face became markedly edematous and swollen, and an eczema-like eruption developed over the entire surface, especially on the thighs, as numerous pustules. The patient had at this time attacks of dyspnea. Morning temperature, 38.5° C.

Aug. 15.—The eyes are about closed by the swelling of the lids, admitting of only slight opening on effort. The mouth stands out like a proboscis and the lips are markedly swollen. On the chin and both cheeks, on the upper lip and in the nasal outlets are honey-yellow crusts; the same on the neck, the right upper extremities, and the upper right portion of the thorax, the inner surface of both thighs, and in less degree on the left upper extremity. Two days later the swelling of the face had markedly subsided; the eyes could be readily opened. Temperature was normal. General condition good.

Aug. 25.—The swelling and redness have completely disappeared, and there remain but a few spots that are still slightly reddish.

[The author evidently believes the naphthol responsible for the outbreak, but it is possible that iodoform may have been the etiological factor.—ED.]

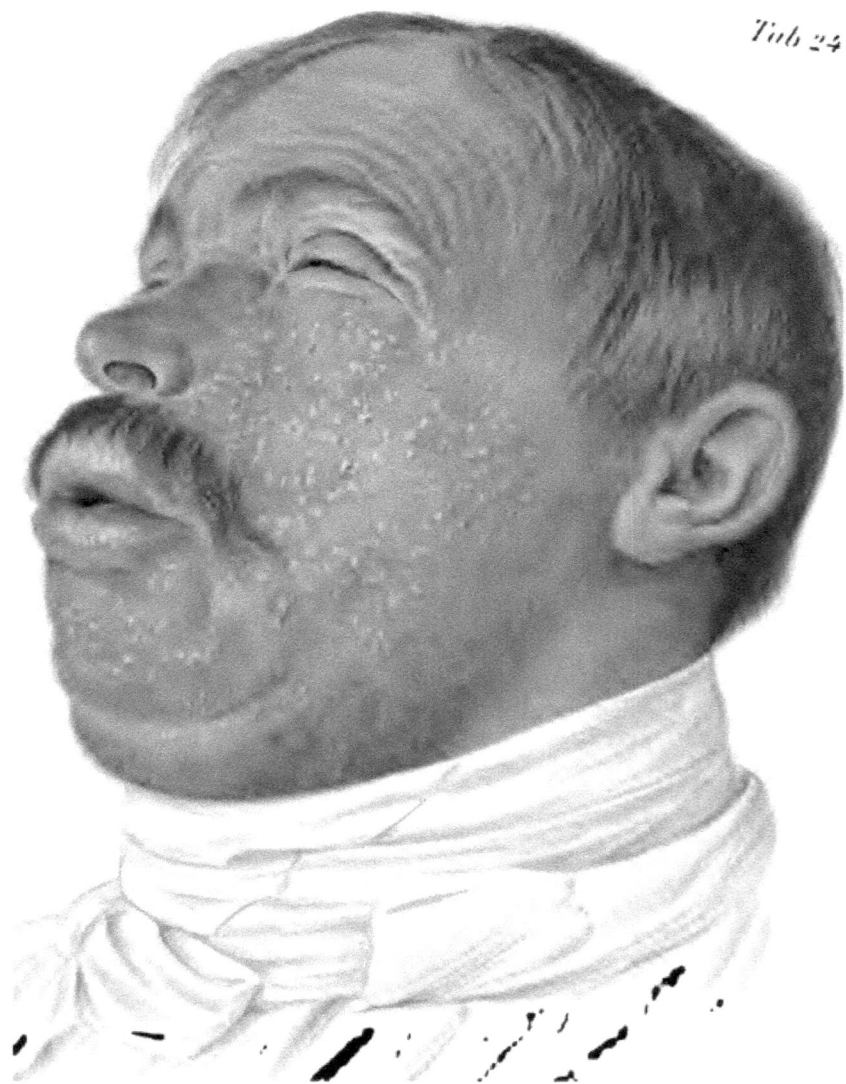

Tab 24

PLATES 25 and 25 a.

Eczema Pustulosum Artificiale [Dermatitis – Ed.].

B. Ph. was admitted for a markedly inflammatory eruption about the legs, which he stated had followed the use of a salve made up of three parts of diachylon ointment and two parts of vaselin. He had applied this to his legs for the relief of an alleged eruption which had been itchy, and had rubbed it in repeatedly with great vigor. After five days' use of this ointment the present eruption made its appearance. Three days later he was admitted to the hospital.

Status Præsens.—The extensor surfaces of both legs to the lower third, also the posterior surface of the right thigh near the knee, are the seat of numerous irregularly-grouped large pustules. Out of the center of each pustule emerge one or more hairs. The skin immediately surrounding the discrete lesions is reddened; where these are in groups this redness is confluent. The color is bright red, and may be made to disappear momentarily by pressure. There is no pronounced infiltration. The oldest of the pustules and purulent blebs show already hemorrhagic contents. Some have been broken and have given place to reddish crusting. The rest of the body is entirely free from efflorescences. Here and there are scratch-marks, especially on the flexor surfaces and at the axillæ.

The patient remained under observation for two weeks, during which period several boils developed; at the end of this time all the pustules had dried up, and from most the crusts had already fallen off; the boils had also practically run their course.

[The case would be classified with us as a follicular pustular dermatitis, which is occasionally noted to follow the vigorous rubbing-in of ointment (especially if not very fresh) on hairy parts.—ED.]

Tab. 25.

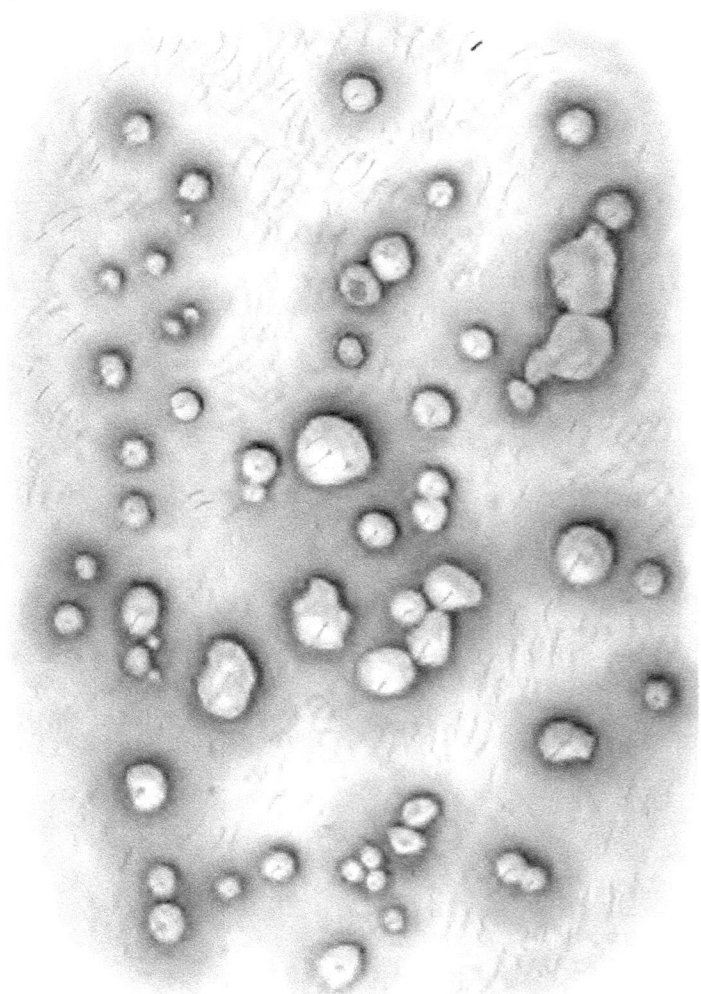

Tab. 25 a.

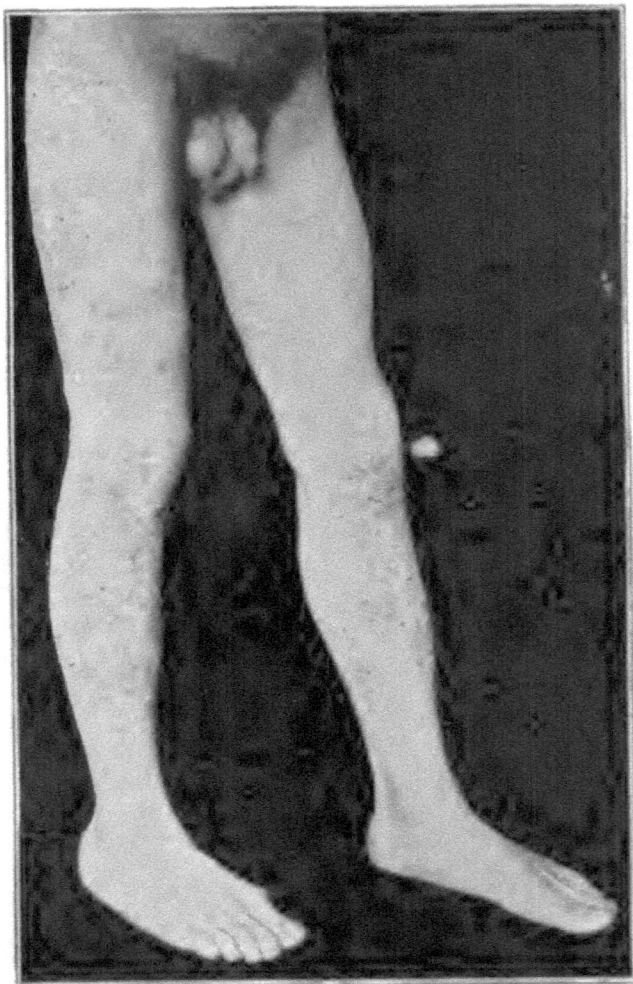

PLATE 26.

Eczema Marginatum (Tinea Trichophytina Cruris).

B. F., aged 15 years, schoolboy, stated that the eruption had first made its appearance several years before, primarily on the anterior surface of the right thigh, and later on the left, in the pubic region and about the genitalia. There had been slight itching.

Status Præsens.—The skin of the middle surface of both thighs, to the inguinal furrow and up to the pubic region, is bright red and hard to the touch. Toward the normal skin the affected area is bounded by a reddish-brown, slightly-scaly, irregular border. The border is elevated and made up of a continuous line of confluent papules, pinhead to lentil in size; the middle of the area is, for the most part, grayish-brown pigmented and slightly rugous. Beyond the main area of disease are a number of characteristic ring-shaped patches. On scrotum and penis are similar efflorescences, but much more recent and ring-shaped. The disease exists in the axillary regions, also, as typical, sharply-defined, scaly, confluent areas.

Treatment.—Lysol lotion (5 per cent.) and washings with soap, naphthol-soft-soap, and applications of Lassar's salve sufficed to cure the patient in thirty-one days.

[It is not now generally believed that all cases similar or closely similar to that here described are due to the ringworm-fungus, but that some may be classed as a variety of eczema seborrhoicum; the large majority, however, undoubtedly belong in the ringworm-group, in which the author has placed this case.—Ed.]

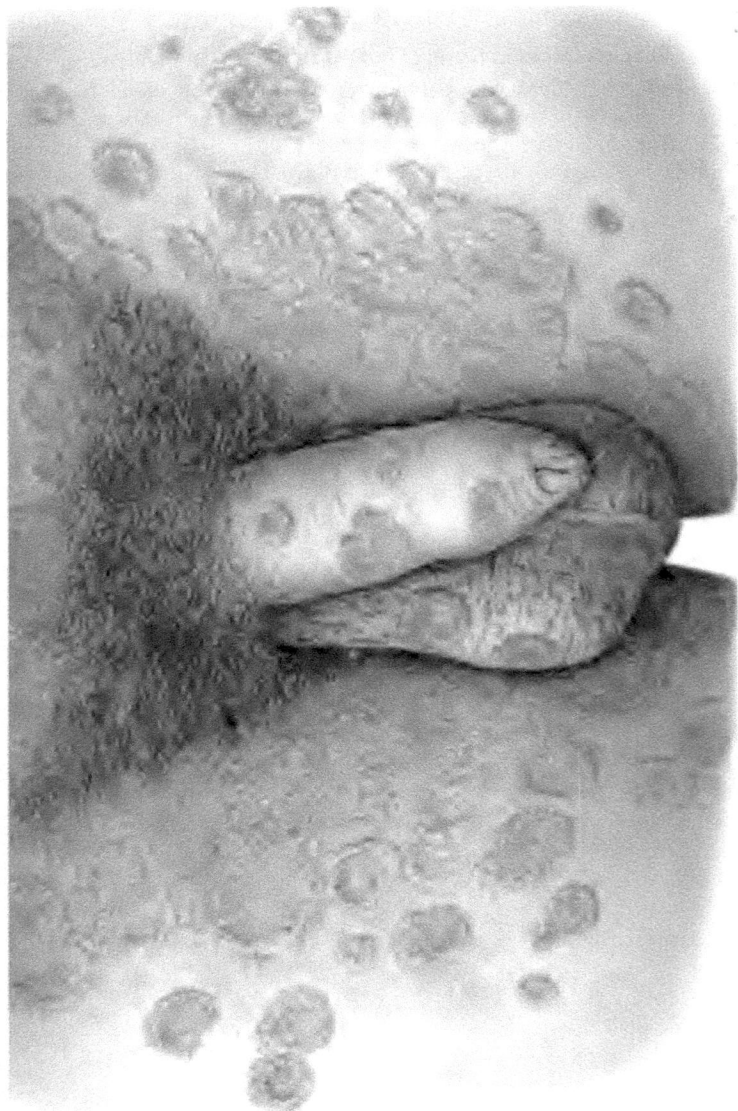

PLATE 27.

Eczema (Mycoticum ?).

N. N., butcher's assistant.

Status Præsens.—Beneath the right nipple is a half-palm-sized crusted area. The crust is of a yellowish-green color, and the border of the patch is red. On this border is seated an almost continuous row of white vesicles and blebs; a portion of the periphery consists of slightly-detached epidermis. Close to, but beyond, the patch are scattered small blebs with red areola.

[This would be considered by some dermatologists as an example of so-called "parasitic eczema," and by others as a "pyo-dermia" or a "pyogenic dermatitis."—ED.]

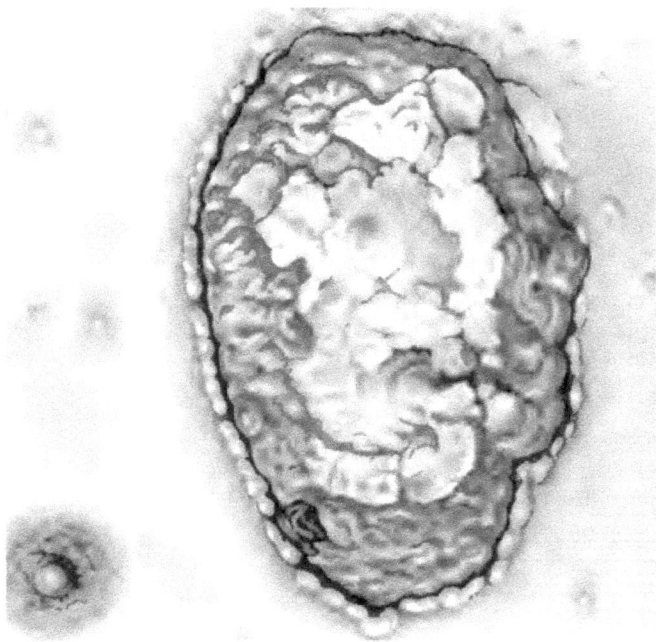

Eczema Madidans et Crustosum (Mycoticum ?).

J. S., aged 28 years, miller, came under observation Oct. 20, 1896. The disease had existed since June, without known cause.

Status Præsens.—On the left lower leg is a palm-sized, irregularly-bounded area, covered with dry yellowish crusts. Upon removal of the crusts the oozing corium is brought to view. In the immediate neighborhood of this patch are a few lentil- to dime-sized pustules. On the left upper extremity, on both the arm and forearm, are similar areas, partly oozing and partly crusted, of the size of a thumb-nail to a silver quarter.

Upon the application of diachylon ointment, and later the application of Lassar's paste, a cure resulted.

[This case is essentially similar to that of Plate 27.—Ed.]

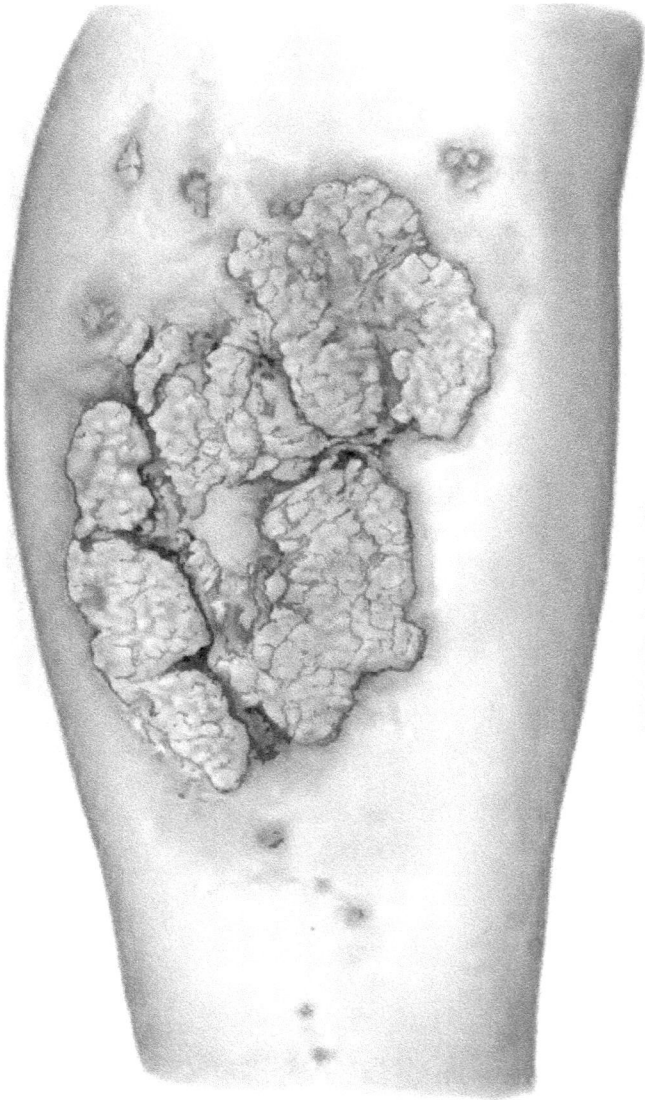

Tab. 28 a.

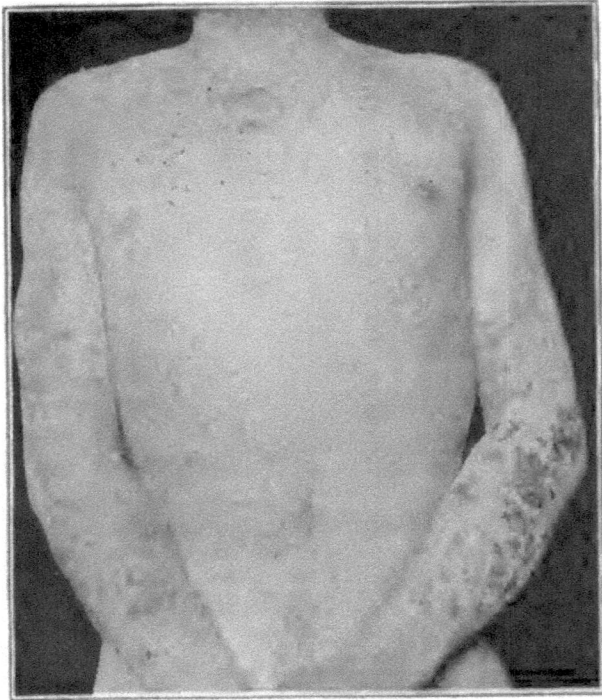

PLATE 29.

Prurigo.

B. K., aged 13 years, school-girl, admitted Aug. 26, 1897, stated that the skin-affection had existed since earliest childhood.

Status Præsens.—The skin of the extensor surfaces of the extremities, especially the lower in the knee-region, is thickened, dry, and rough, the folds exaggerated and of a brownish color. There are numerous embedded papules, many covered with blood-crusts; between these are reddish and brownish pigmented spots, the sites of former lesions.

Under treatment with macerating baths and the application of salves the condition was somewhat improved, and the patient left the hospital after twenty-six days' treatment.

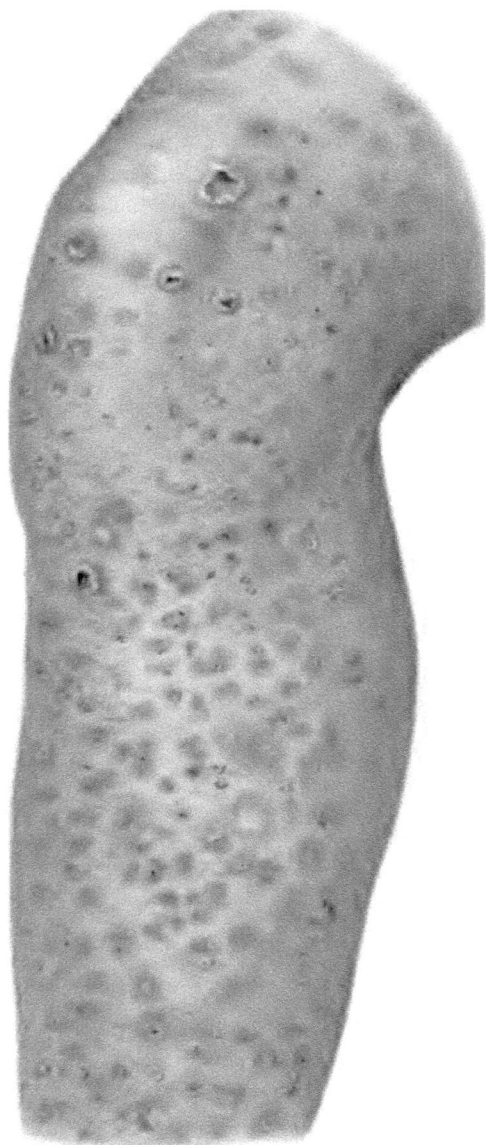

PLATE 30.

Acne.

M. A., aged 17 years, somewhat pale, had suffered for several years from continued outbreaks of papules and pustules on the face, neck, shoulders, and back, having their seat in the sebaceous glands and ducts. The lesions vary in size from a pinhead to a lentil, the larger lesions showing in the central part purulent contents.

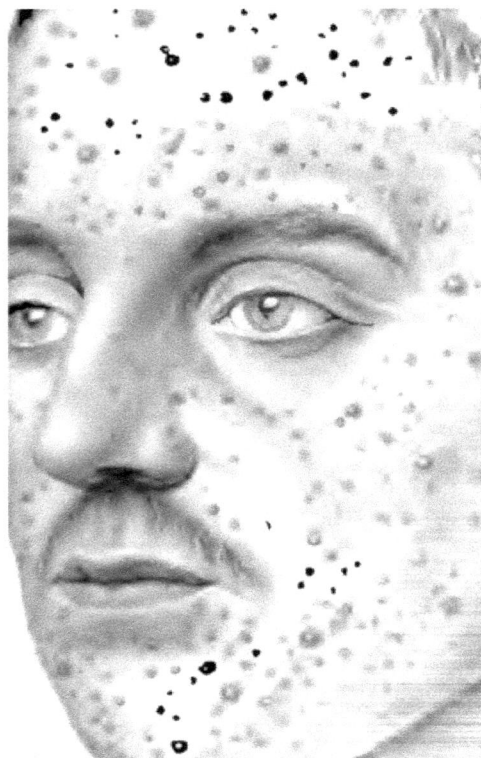

PLATE 31.

Sycosis.

T. A., aged 54 years, admitted May 7, 1896, stated that in 1890 he noticed a papule on the upper lip, which later changed into an oozing spot, while immediately in the neighborhood other papules and oozing patches appeared. In the course of four years the process had spread and gradually involved the entire upper lip.

Status Præsens.—The entire moustache-region is inflamed and crusted, the crusts mostly confluent and of a greenish color; here and there a small spot covered with horny epidermis, and in some places moist spots with a papillomatous (frambesiform) tendency. The crusting is quite thick in places, and is irregularly and scantily pierced by projecting hairs, which are loosely embedded and which may be quite easily and painlessly pulled out, the root-sheath appearing swollen. In the middle of the lip toward the right upper part the skin is free from crust, reddened, and infiltrated, and to some extent covered with scales. There is no active purulent discharge. From the corner of the mouth the process tends to extend downward, although the crusting here is not so massive and is easily detached.

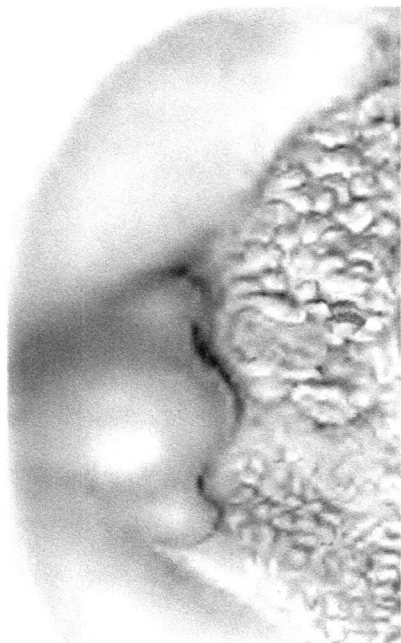

PLATE 32.

Furunculosis [Ecthyma, Impetigo Contagiosa?—Ed.].

T. J., aged 36 years, hostler, was admitted Apr. 1, 1897. The skin-affection had existed eight days. The first efflorescences appeared in the region of the coccyx and then spread toward the sacral region and the lower extremities. In the past three days the patient had had several chills.

Status Præsens.—The patient is large, strongly built, and well nourished. The buttocks and the outer side of the left thigh, and to a much less extent on the remaining parts of the lower extremities, except the extensor surfaces, are seen dime-to dollar-sized red crusted lesions. The crusts are of a brownish-yellow color, somewhat heaped up, and are firmly adherent to the underlying superficial ulceration. The base is inflammatory and infiltrated, and the areola pronounced and also inflamed and infiltrated. The process is somewhat deep, having almost a furuncular nature. Cultures made from the secretion show staphylococci and streptococci.

Under antiseptic applications a cure resulted in sixteen days.

[These lesions, as here depicted, correspond to the lesions usually described under the name of ecthyma. By some observers they would also be looked upon as a markedly inflammatory type of impetigo contagiosa. In fact, at the present day there is a growing belief that these various diseases are the same in etiology (pus-cocci), the differences in the objective phenomena being due to individual peculiarity of the skin or variation in its resisting power.—Ed.]

Tab. 32

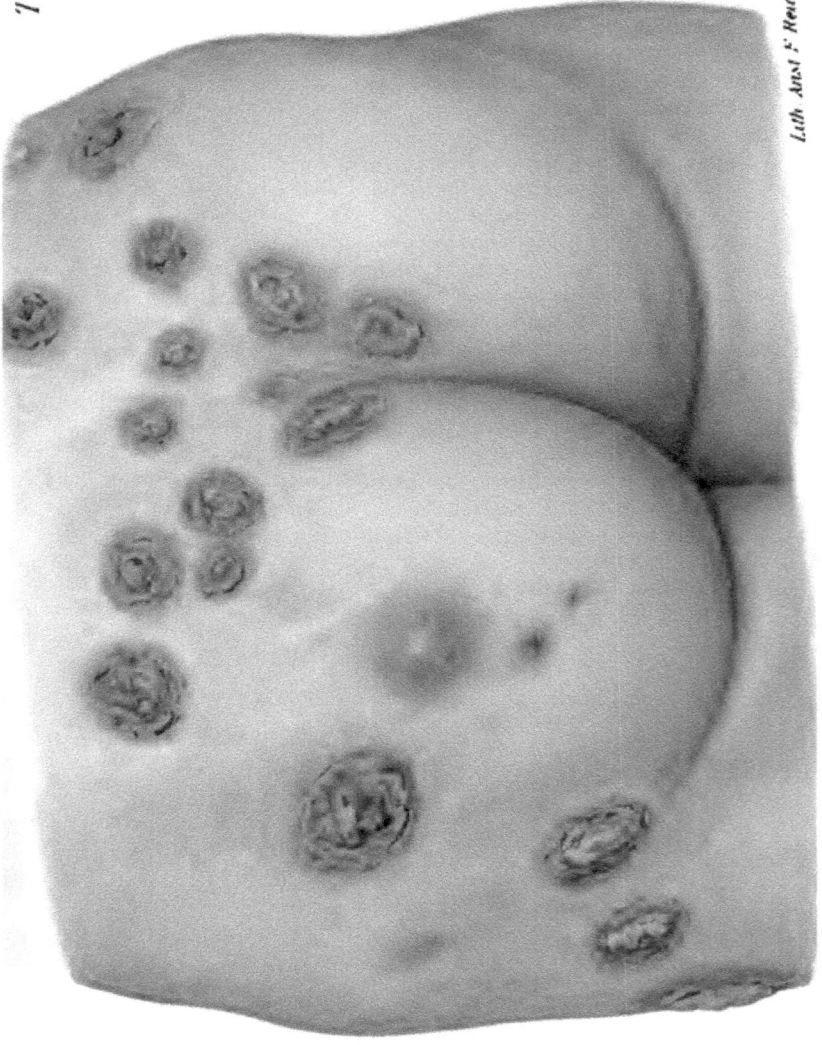

Lith. Anst. v. Reichhold Munc.

PLATES 33, 34, and 34 a.

Pemphigus Vegetans.

H. O., aged 78 years, female, was admitted Feb. 28, 1894. The patient was in the hospital three years previously for a pemphigoid eruption. The present attack began three weeks ago.

Status Præsens.—The nutrition of the patient is poor; the hands and feet slightly edematous. The entire surface is the seat of a bleb-eruption; some of the older lesions covered with crusts. Immediately surrounding the anal outlet are some papillary growths. The skin of the neck, back, axillæ, and genital region is considerably pigmented, without recognizable cause.

Course.—Up to August of the next year (1895) the patient was upon two occasions permitted to leave the hospital, inasmuch as she was free from blebs and felt much better as to general health. Since Aug., 1895, however, she has been constantly in the hospital. During this whole period it suffices to state that the entire body was the seat of recurrent outbreaks, of which the following description is a picture: The yellowish-brown pigment of the earliest period had, on the neck, upper shoulders, lower abdominal region, and axillæ, changed to a dark-brown or blackish tint. The skin of the hands and inner thighs felt leathery and was more or less rugous. In the axillæ are flat warty and papillomatous thickenings of the skin; these areas had formerly been moist, deeply furrowed, and papillomatous, and coated with a cheesy covering. In the genitocrural folds, on the labia, and surrounding the anus are red papillomatous growths, seated upon broad bases and discharging a cheesy secretion. On the dorsal surfaces of the hands are fresh pemphigus-blebs and abraded areas, the seat of recent lesions. The skin is furrowed and leathery, and the border of the blebs red and swollen. On the face and lips are smaller broken blebs. On other parts of the body may be seen small blebs, associated with troublesome itching. The patient is considerably emaciated and depressed in general health. Since the beginning of the present year the condition has measurably improved; the bleb-outbreaks have lessened both in extent and severity, and the subjective sensations are not troublesome.

Treatment.—Great care was taken as to cleanliness; the papillomatous excrescences were treated with drying-powders and lotions. Internally, in addition to tonics and nutritious food, arsenic was given about two months and "Brown-Séquard" for three-quarters of a year. At present writing nothing is being administered.

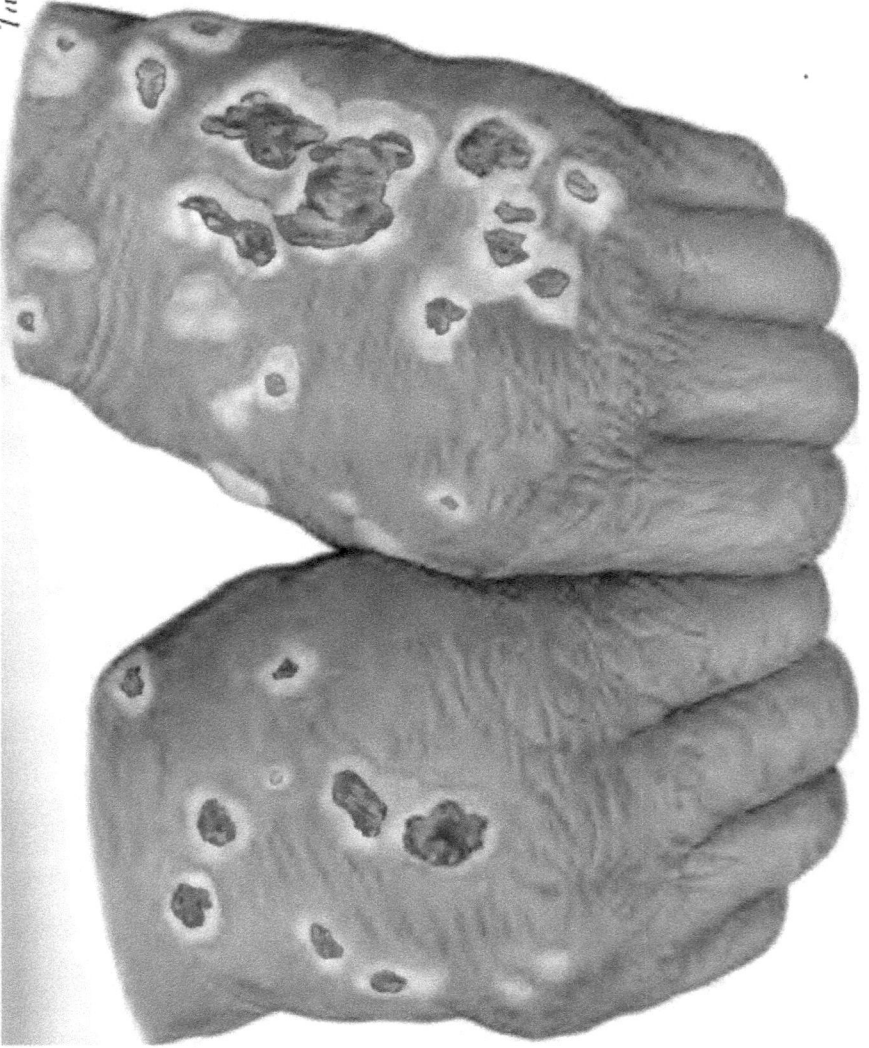

Tab. 14.

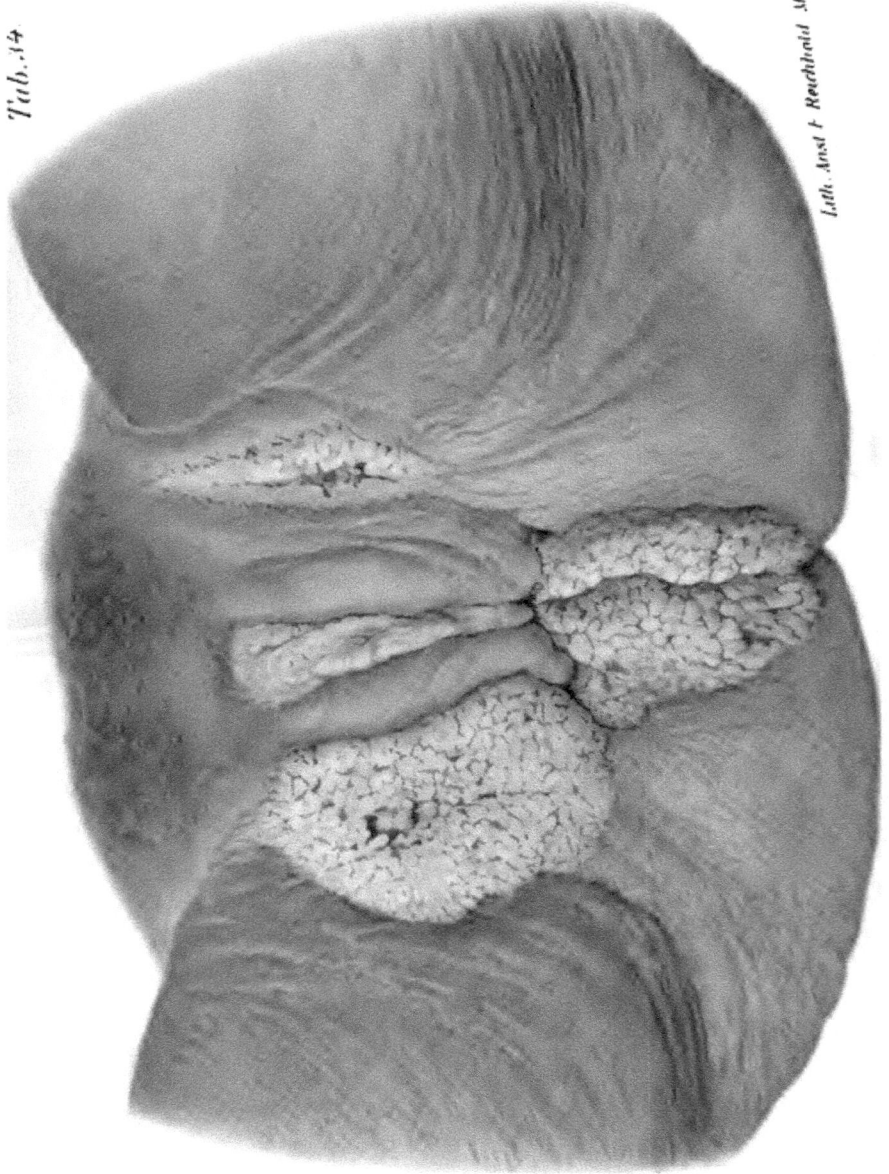

Lith. Anst. v. Reichhold, München.

Tab. 34 a.

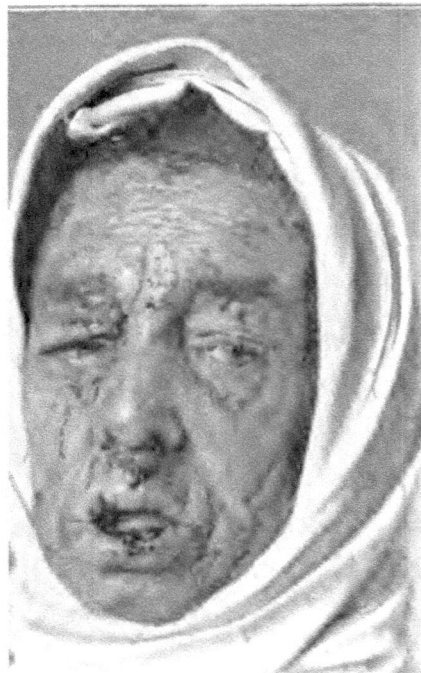

PLATE 35.

Nævus Verrucosus.

M. H., aged 27 years, female.

Status Præsens.—The patient had between the shoulders an elongated, oval, brown-pigmented patch. The periphery is of a light-brown color and slightly elevated; the center is dark brown, with smooth, rounded, wart-like projections which feel somewhat elastic. No pain or tenderness.

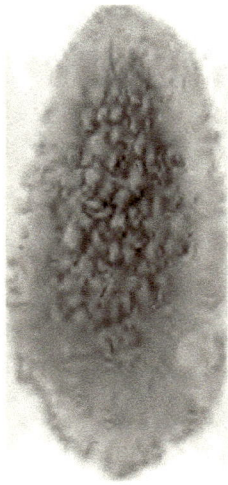

PLATE 36.

Nævus Pigmentosus Unilateralis.

P. C., aged 22 years, male, had shown since infancy pigment-marks; they had not caused any annoyance.

Status Præsens.—The skin over the buttock, from the anal furrow toward the right and downward on the thigh, irregularly bounded, is of a yellowish-brown color; otherwise of normal structure and sensation. Further down, beginning at the lowest part of the thigh and extending to the posterior and inner surface of the lower leg down to the foot was a similar blemish.

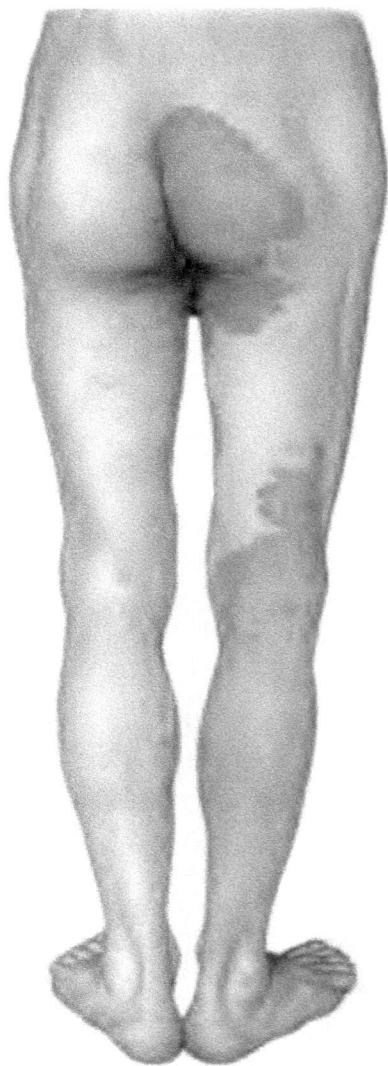

PLATE 37.

Hyperchromatosis Arsenicalis.

L. F., aged 24 years, male, stated that in July, 1895, for four weeks, and from late August into October, for six weeks, he had been in the hospital for the treatment of a scaly skin-eruption. Both times treatment was begun with the administration of five drops of Fowler's solution daily, and reached in the first course of treatment twenty drops and in the second period twenty-five drops. The scaliness had gradually disappeared and given place to extensive pigmentation.

Status Præsens.—The hairs on the patient are black and the skin yellowish-brown. On the extensor aspects of the fore-arms, elbows, and knee-regions are numerous scattered psoria-sis-efflorescences, partly covered with scales. The skin of the entire body, with the exception of the face, neck, hands, and feet, is the seat of sepia-colored, reddish-brown spots and areas. The most are discrete, pea- to half-dollar-sized, although there were many confluent areas of larger size and irregular shape, melanotic in character. In most of the discrete spots the central part is less dark, and the borders gradually merge into the surrounding normal-colored skin. Neither scaliness nor swelling of the skin is noticed; the melanotic spots and areas feel normal to the touch and show the normal lines. The patient had during his two arsenical courses 340 and 570 drops respectively, or, in all, 900 drops.

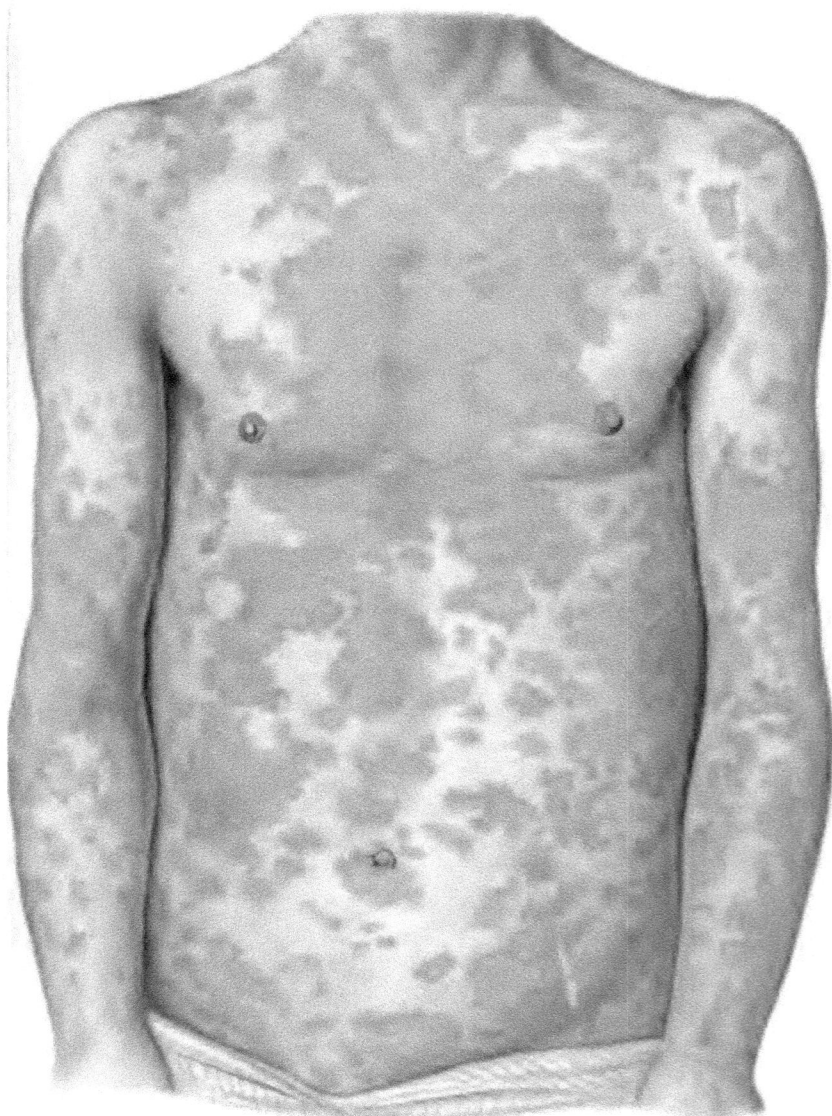

Tab.37.

PLATE 38.

Lichen Pilaris (Keratosis Pilaris).

J. H., aged 18 years, female, stated that she has had a rough, hard skin for some years. Recently she observed the appearance of numberless minute brownish points.

Status Præsens.—Patient is strongly built and well nourished. The entire skin shows want of care. The extensor surfaces and the back are the seat of numerous, irregularly-scattered, pinhead-sized, brownish-colored papules, having their seat at the hair-follicles. The skin feels rough and dry, more noticeable on the extensor surfaces of the extremities. There was no itching, nor any symptom of a subjective character.

[In most cases of keratosis pilaris, as observed in this country, the manifestation is most marked on the thighs, especially the anterior and outer aspects; in fact, it is seldom that parts other than the thighs and corresponding surfaces of the arms and forearms are perceptibly involved.—Ed.]

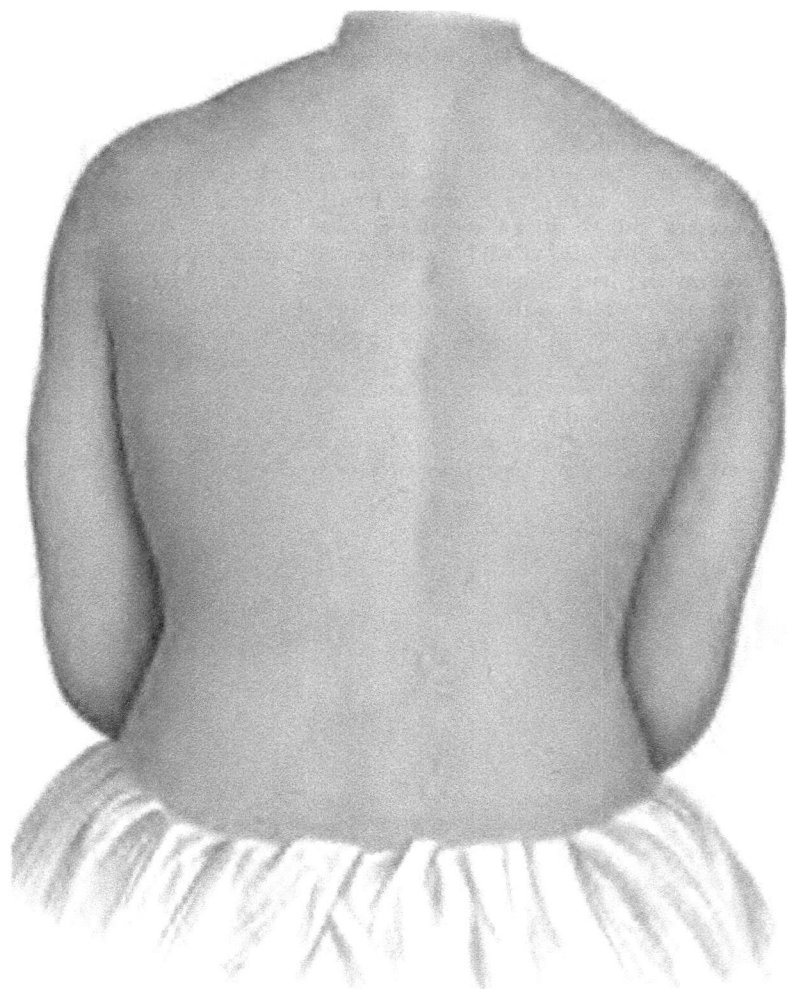

PLATE 39.

Ichthyosis.

A. K., aged 27 years, female, came under observation Mar. 29, 1897.

History.—The roughness and cracked condition of the skin had existed since infancy. Her only brother was also the subject of the disease.

Status Præsens.—The patient is medium-sized, and moderately nourished, but pale. The skin of the entire surface, especially the abdomen, back, and lumbar region, is rough and covered with epidermic lamellæ and plates, the cracks and fissures dividing the plates and scales disclosing the reddish rete.

[As a rule, there is less of the red aspect in ichthyosis-cases than is here pictured, and in most instances it is entirely lacking. Occasionally, however, especially when an eczema is superadded, as sometimes happens, the hyperemic element is conspicuous.—ED.]

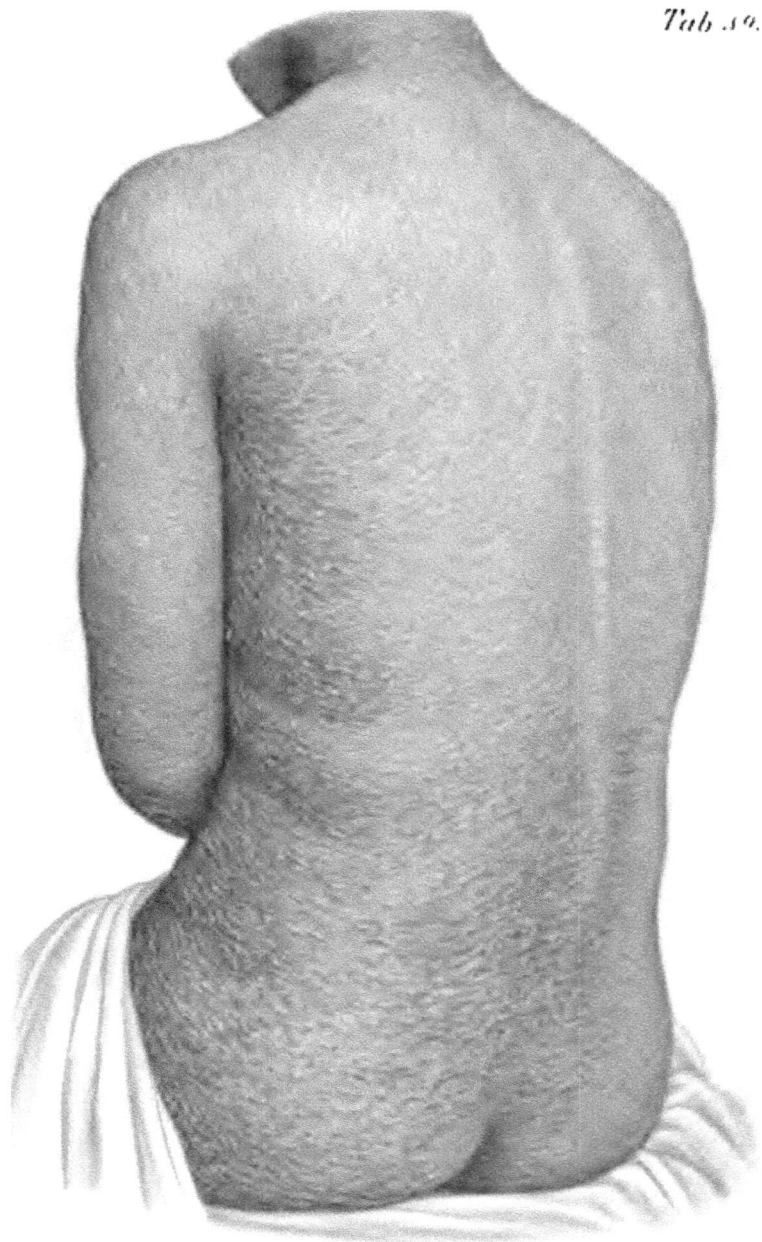

Tab 80.

PLATE 40.

Hyperkeratosis Palmaris (Callositas).

L. K., aged 36 years, day-laborer, was admitted Nov. 20, 1896. Patient was a digger, and believed his occupation responsible for his complaint; he had at an earlier period had the ordinary callous areas in the hand. The present condition, it was stated, had lasted two weeks. The patient had long suffered from foot- and hand-sweating.

Status Præsens.—On the palms and flexor surface of the fingers of both hands, but more especially the right, the skin is much thickened, the epidermic accumulation consisting of many layers. The greatest thickness is to be observed on the thumb, the ends of the first, second, and third fingers, and on those places against which the handle of his shovel had pressed most. The joints of the parts showed tolerably deep cracks and fissures. The patient held the fingers of the right hand in a flexed position, and experienced pain in attempts to straighten them out. The nails were likewise much thickened, and between the matrix and nail was a mass of hardened epidermic accumulation. Similar conditions, but in much less degree, were noticeable on the soles of the feet.

Treatment.—Hand-baths, soft soap, and diachylon salve. Patient was discharged cured at the end of seven weeks.

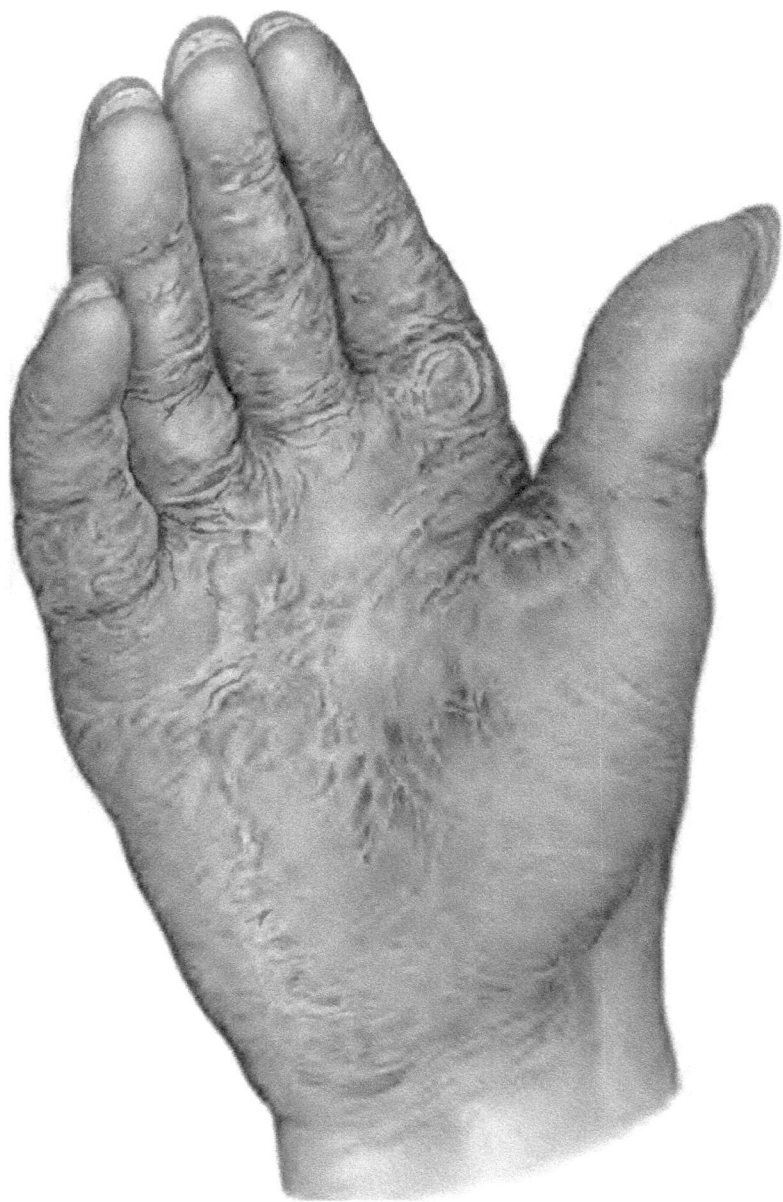

PLATE 41.

Leucoderma (Vitiligo).

Z. D., aged 21 years, washerwoman, of dark complexion and dark hair.

Status Præsens.—The skin of the inner sides of the thighs, the groins, the labia, and perineum is wanting in pigment-matter, being of a dead-white color; there is increased pigmentation in the surrounding skin. The hair on the labia and pubic region is, for the most part, also white. With the exception of the whitening of the skin and hair there is nothing abnormal.

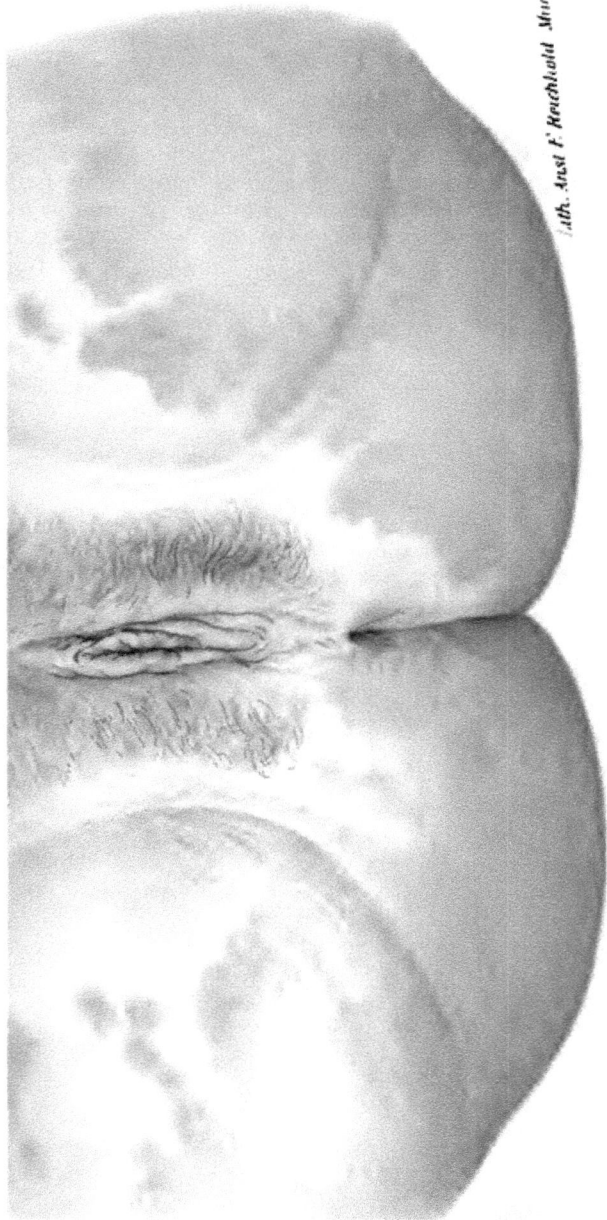

PLATE 41 a.

Alopecia Areata (Alopecia Totalis Neurotica).

N. N., aged 22 years, female, unmarried, was admitted Oct. 13, 1896.

History.—Patient was of a highly-nervous, excitable family No one had, however, previously suffered from any hair-loss. As a child she had varicella, and later, in her fifth year, diphtheria. Since that time she had remained anemic and weakly, and seemed unable to regain her former condition of health. As a young girl she had light-blond, very luxuriant, long hair. In early childhood she had suffered from a seborrheic condition of the scalp, but this was not accompanied by any hair-loss. From her seventh year she had suffered much from periodical one-sided headache, which, for the most part, was worst toward the occiput and neck. She began to menstruate when eleven years old, at its first onset being under great nervous perturbation; since then she had menstruated, without any special difficulty, regularly every three weeks. Some months after the establishment of this function she was subject to severe migraine, since which time she had noticed that her hair was becoming somewhat lighter in color, hard, brittle, and split at the ends, appearing as if without life. After persistent headache and recurrent nose-bleed the patient was brought in an unconscious condition to the nervous clinic. At that time she is said to have been delirious, boisterous, confused in her talk, and to have had convulsions. In one night she lost all the hair of the scalp, axillæ, mons veneris, eyebrows, and eyelashes, and later the downy hairs as well. When the patient recovered consciousness and left the hospital, three weeks after admission, she was completely bald, and remained so for ten years, up to the end of 1894. The nervous symptoms, it was stated, disappeared at this time. She noticed that the scalp-skin seemed firmly attached to the underlying tissue.

In the next six months, up to the spring of 1895, there appeared in places, first in the occipital region, then on the vertex, and finally on the parietal regions, a scanty supply of hair. This grew in length to the shoulders, although it remained sparse in quantity. With return of the severe migraine and nervous excitability the hair again fell out as before in two to three weeks. In another interval of freedom from nervous symptoms, the past five months, the hair now present had grown; the past seven or eight weeks a downy growth had also shown itself in the axillæ and the genital region.

Status Præsens.—By general examination nothing is found except a blennorrhagia of the vagina and uterus. The sensibility, pressure-, pain-, and temperature-sensations are normal, except a slight disturbance in the region of the frontal branch of the facial nerve. The skin-, muscle-, and tendon-reflexes are present. Urine-examination gives a marked increase in phosphates. The nails are milky and show lines running lengthwise, and nail-ends tending to be fragile; they are white-dotted here and there. The skin of the scalp is pale, smooth, shining, and movable upon the underlying part, although not readily lifted in folds. The hairs are thin and atrophic, the longest being six to eight inches, and the lanugo-hair one or two lines long. Some parts, well defined and tolerably symmetric, are almost completely bald, and these, as well as those now covered with hair, agree in their arrangement with the distribution of the skin-nerves (ramus prim. trigemini, II. and III. nerv. cervicalis). A few hairs are on the region of the eyebrows; the lashes are almost completely wanting. The entire skin-covering, especially of the extremities, is dry, in spite of the fact that the patient for a number of weeks has had considerable sweating with the attacks of headache.

Noteworthy is the coincidence of the rapid hair-fall with the psychosis; the occurrence of nervous symptoms, as migraine, congestions, nose-bleed, with the oscillation in growth and falling out of the hair; the symmetry of both the hairy and the non-hairy areas, the distribution corresponding to the skin-nerves; the trophoneurotic disturbances of the nails; and also the hereditary nervous tendencies. The entire course of the affection spoke for the nervous origin of the hair-loss, and the case is to be placed in the class of alopecia totalis praematura neurotica.

PLATE 41b.

Alopecia Areata; Canities.

G. P., aged 17 years, salesman, stated that in Feb., 1895, he suffered from alopecia, which by spring was entirely cured.

The present affection appeared in Jan., 1896; the hair changed to a white color in two places in the occipital region and then began gradually to fall out. On the borders of the circular areas the hairs are easily pulled out.

Tab. 41 a u. b.

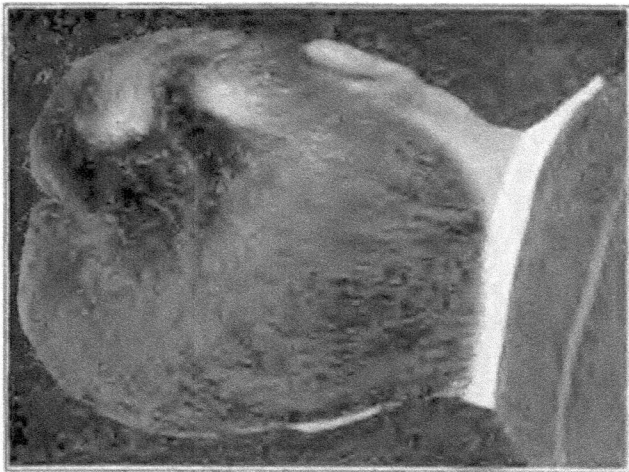

Tab. 41 b.

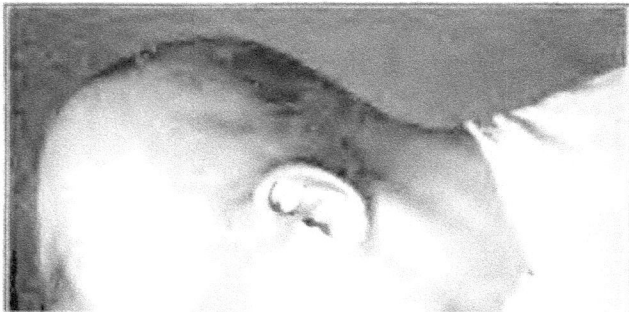

Tab. 41 a.

PLATE 42.

Lupus Erythematosus.

F. H., aged 38 years, female, noticed for a number of weeks the appearance of an eruption on the end of her nose.

Status Præsens.—The disease is seated upon the tip and left ala of the nose. The area is slightly elevated, and is sharply defined from the healthy skin by a red, somewhat raised border. There is slight scaliness, the scales being markedly adherent and of a grayish and greenish-gray color; upon their removal the base is noted to be livid red.

A salicylated mercurial plaster was advised, under which the patient was improving, when she failed to continue her visits to the dispensary.

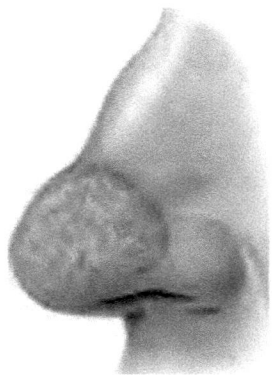

PLATE 43.

Lupus Erythematosus.

N. E., aged 40 years, admitted Apr. 15, 1898. The disease was first noticed two years ago, appearing on the nose and near by on the cheeks. Under treatment improvement then ensued. The present exacerbation patient observed eight weeks before admission.

Status Præsens.—Nose, cheeks, and ear show patches in various stages of the disease. The areas are all slightly thickened and red. In some the surface appears stretched and shiny. For the most part, however, the patches are covered with lightly-adherent whitish scales, somewhat greasy in character.

Treatment.—Improvement ensued from a salicylated mercurial plaster; the patient was discharged, and subsequently treated in the dispensary.

Tab. 43.

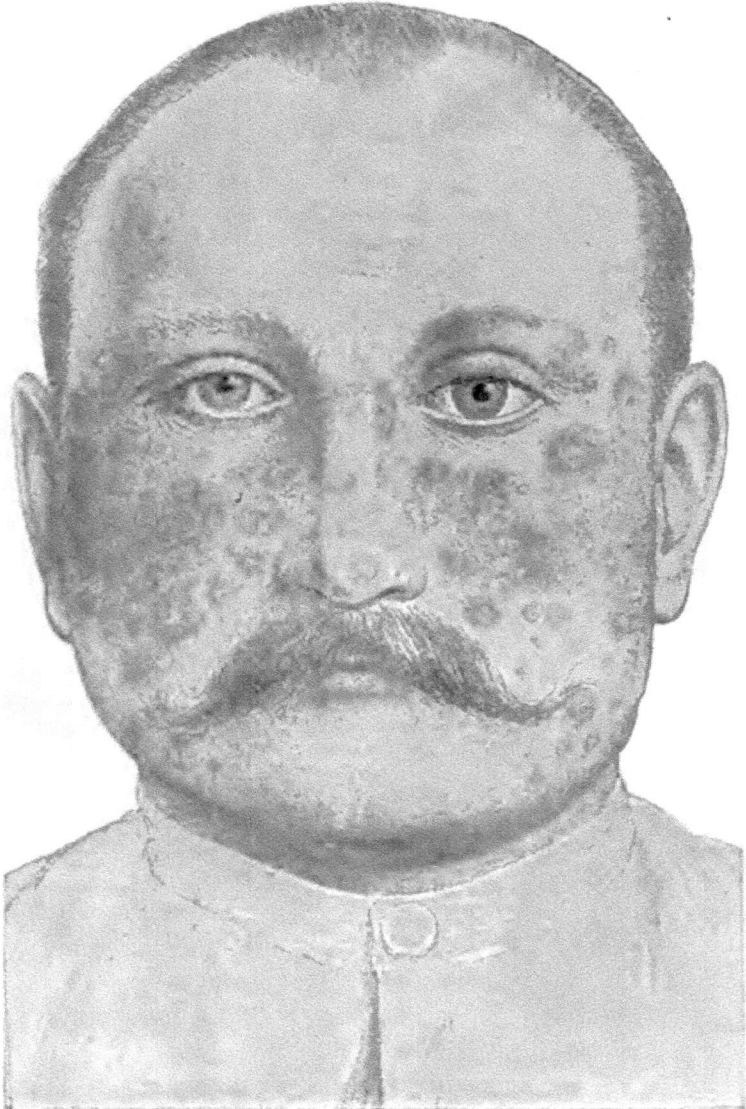

PLATES 44 and 44a.

Xanthoma Tuberosum.

R. P., aged 42 years.

History.—The father of the patient died of liver-disease; his mother and brothers and sisters are living and healthy. About ten years ago the patient began to notice the appearance of small tumors on the extensor surface of the upper extremities. They gave rise to no discomfort, except when struck or pressed upon, when they felt slightly painful. In the course of the year similar growths made their appearance on the nape of the neck, on the buttocks, and on the extensor surfaces of the lower extremities. For the past three years the condition has remained about stationary.

Status Praesens.—Patient is of medium size, strong, but pale, with considerable panniculus adiposus. The internal organs are normal. On the nape of the neck near the border of the hair, on the extensor surfaces of the upper extremities, the buttocks, and the extensor surfaces of the lower extremities, the skin is the seat of numerous growths, lentil to hazelnut in size, rounded and prominent; partly smooth and partly cleft. The borders are of a bright-red color; toward the center of the growths this becomes of a fat-yellow color. Between the closely-crowded and confluent growths are lentil-sized cicatricial depressions, with an irregularly-pigmented border.

Urine-examination: Albumin, about 0.067%; sugar, 5%; quantity passed in twenty-four hours, 1200 grams; specific gravity, 1031; color, wine-yellow, clear; no renal elements in sediment.

Histologic examination of one of the growths shows that the tumor consists of fibrous filaments taking origin out of the connective tissue of the skin; the yellow fat lies in the cells of the fibrous filaments. The tumor is not inflammatory in origin.

Treatment.—Patient was advised to take a six weeks' course at Carlsbad, with restriction of albuminous foods and the use of abundant vegetable food. During this treatment there was a remarkably rapid involution of the xanthoma-lesions; in such growths the peculiar scar-like depressions with pigmented areola, above mentioned, remained; the sugar disappeared entirely from the urine.

The patient presented himself at the end of May, 1898, with new nodules; and sugar had reappeared in the urine.

[This case was published in full by Dr. G. Toepfer, in the *Archiv für Dermatologie und Syphilis*, Band 40, 1897.]

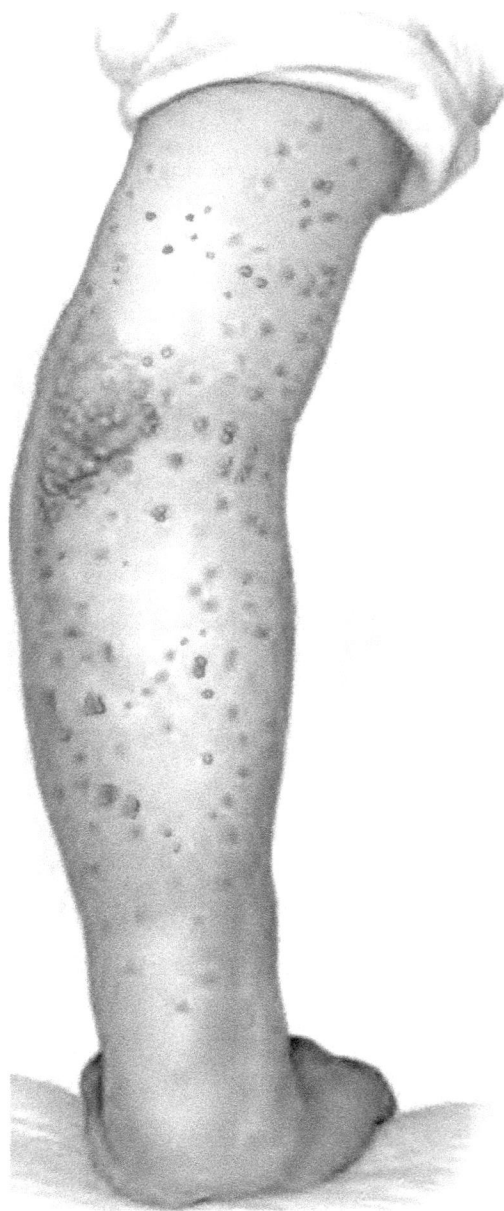

Tab. 14 a.

PLATE 45.

Nævus Vasculosus.

.

PLATE 45 a.

Nævus Vasculosus et Verrucosus.

The seat of the affection, the left temporal region, is **copper-
red** from minute capillary enlargements, and irregular in
shape. Scattered **over it are larger** dilated capillaries, and
small growths **or elevations** made up of hypertrophic connective
tissue and blood-vessels.

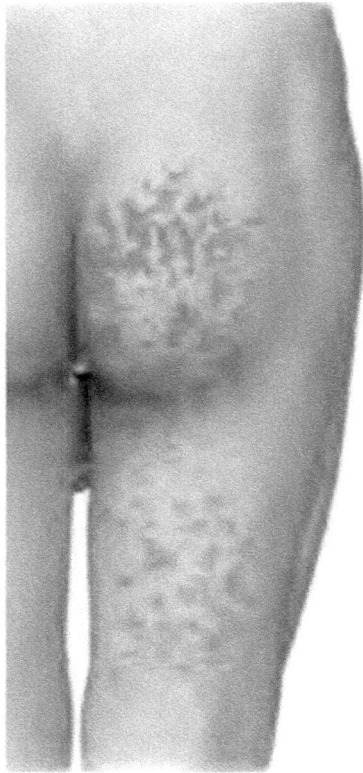

Tab. 45 a.

PLATE 46.

Lupus Vulgaris (Lupus Serpiginosus).

K. A., aged 14 years, female, admitted Mar. 21, 1898, stated that the disease had existed since childhood. Several of her sisters died in infancy; she herself was always healthy.

Status Præsens.—Patient is large, slenderly built, and pale. Apex of left lung infiltrated; heart normal. In the region of the left thigh, involving the upper two-thirds, outer side, is observed a wrinkled scar. In this scarred area are numerous flat, reddish, irregularly-distributed and -arranged tubercles, in greater number and more crowded toward the posterior border. Some of these tubercles are covered with a thin crust; some are redder in color and show minute blood-points. The posterior boundary-line is made up of a thick wall of crust-formation.

Treatment.—Under chloroform, enucleation, Paquelin cautery, and excision. Healing ensued and patient was discharged cured forty-three days after admission.

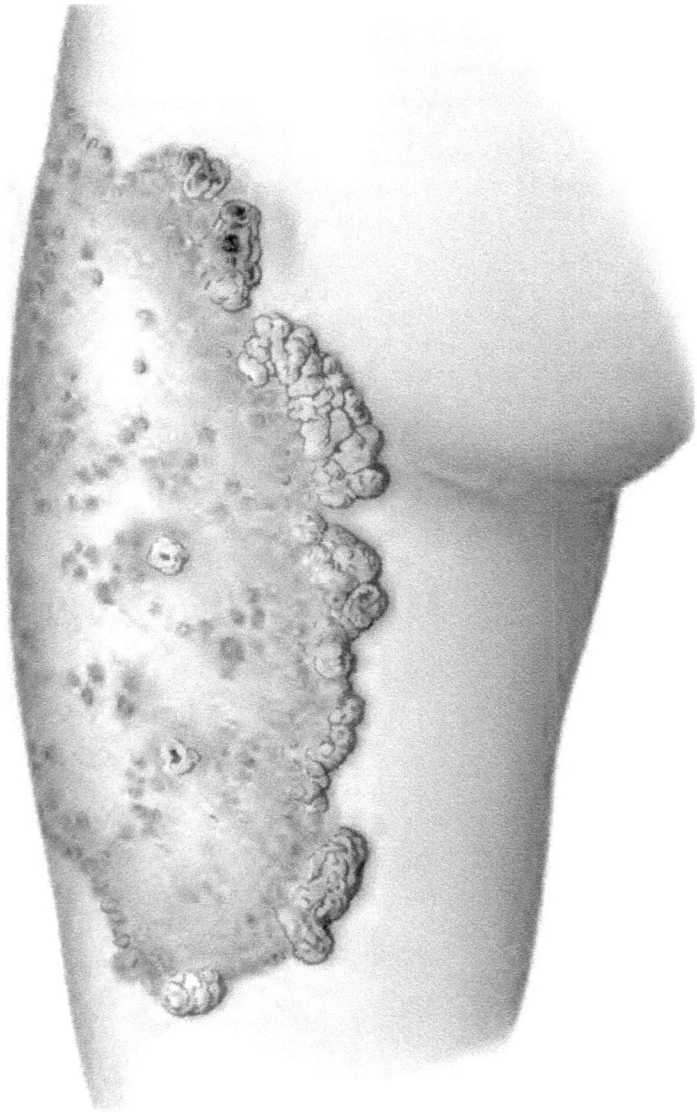

PLATE 47.

Lupus Vulgaris (Lupus Exulcerans, Lupus Exedens).

W. A., aged 22 years, shoemaker, was admitted Jan. 6, 1897. The patient stated that the skin-affection had existed since he was two years old, and that he had been under treatment several times. At present there were some pain and itching. One brother and his parents had died of consumption.

Status Præsens.—On the inner aspect of the left thigh there is a palm-sized, bright-red, infiltrated patch, partly scaly and the central portion cicatricial. On the borders reddish-brown tubercles are to be seen. In addition to this area isolated patches covered with crusts exist near by. After removing the crusts superficial ulcerations are disclosed. On the extensor aspect of the thigh, at the same height, is a half-dollar-sized area, similar in character. Besides these areas there are elongated, atrophic, somewhat depressed scars in the popliteal region.

Treatment.—Under chloroform the patch was excised and the skin stretched from the two sides and stitched together; later the uncovered wound was covered with transplanted skin. The patient was discharged cured four months after admission.

PLATE 47a.

Chronic Tuberculosis of the Hand, following Exarticulation of Necrosed Middle Finger.

K. J., aged 62 years, day-laborer.

History.—The patient was taken sick in May, 1891, four years previously, with severe pain in the middle finger of the right hand; in Feb., 1892, it was found necessary to remove it by enucleation; the parts were necrosed. A year and a half ago the dorsum manus at the region of enucleation showed signs of inflammation and became ulcerated. On the right cheek the patient has had for three years a dollar-sized patch which has changed but little. Five children of the patient are living and healthy. The patient has been in attendance at the dispensary for twelve weeks.

Status Præsens.—Just outside of the right corner of the mouth, separated from it by a narrow piece of sound skin, is a patch of disease about an inch square. The patch is chiefly cicatricial, the scar being white; toward the angle of the mouth the skin is reddened, and in the middle part depressed and slightly ulcerated; papillomatous peripherally. The base of this area is moderately inflamed, infiltrated, and elastic. Toward the lower part of the face just below this area is a small patch with tubercles and ulceration.

The right upper extremity, especially the forearm, is, in comparison with the left, emaciated. The hand itself is slightly bloated, the thumb free, three fingers flexed; extension of the latter is impossible. The middle finger is wanting. Extending over the site of the exarticulated finger toward the palm, and to a small distance on the dorsum and down the first and third fingers, the tissue is of a lived, red color, slightly inflamed and infiltrated. The entire surface of this region is beset with millet-seed- to lentil-sized ulcerations, many of which are confluent and form irregular ulcerated islets or areas, extending down into the corium. On the peripheral portion of this area can be noticed fresh groups of small tubercles, some of which have broken down and formed ulcerations similar to those described. Upward on the hand, beyond this active area, the skin is slightly shining and red, with scattered erosions and with slight infiltration. At the wrist is another area of active tubercles and ulceration.

Oct. 18.—The suppurative tendency in the tubercles continues.

Dec. 3.—Over the joint the small tubercles have disappeared and the ulcerations are nearly healed.

Tab. 47 a.

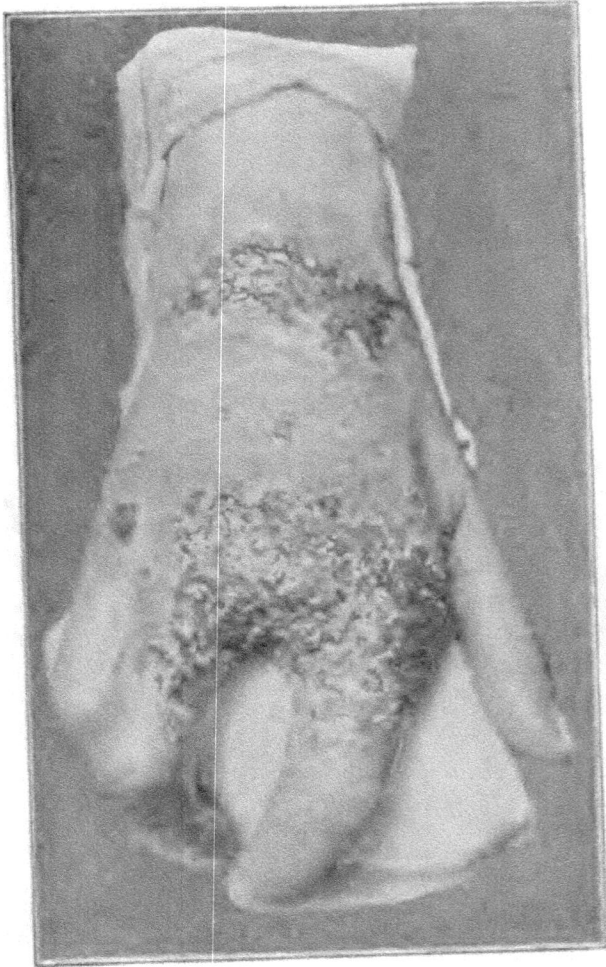

PLATE 47 b.

Chronic Tuberculous Ulcerations on Back of Hand. Scrofulo-gummata on Forearm.

M. M., aged 69 years, female, was admitted Oct. 18, 1895.

History.—The patient stated that in her youth she had been subject to a cough for a long time, which her physician had declared to be a lung-disease. For some years the symptoms of lung-disease had practically disappeared. Ten years ago the patient had caries of the right middle finger, which at first improved, but which two years ago became so much worse that enucleation was practised.

Status Præsens.—The patient is rather slenderly built, but is moderately well nourished. Her muscles are flabby and her skin pale. At apex of the right lung there is a somewhat shorter percussion-sound; some emphysema. The right upper extremity, as to size and nutrition, showed no difference from the left, except that the right middle finger is gone. The scar from the latter reaches considerably up the hand. The surface over the metacarpal bones of the second and fourth fingers, and extending slightly over that of the little finger, is rugous and covered with honey-like crusts, beneath which are shallow ulcerations; the surrounding skin is reddened. On both the corresponding fingers are small tubercles somewhat scaly. On the forearm below the elbow is a livid node about half the size of a hazelnut. Above this, separated by a band of sound skin, is an infiltrated group of similar, but smaller, lesions. Under the olecranon, on the extensor aspect, is a crusted ulcer two-fifths inch wide and over a half inch long, with moderately-inflamed areola, covered with crusts. In the axillæ are several bean- to walnut-sized infiltrated glands.

Treatment.—After removal of the crusts various salves spread upon bandages were from time to time applied.

Dec. 12, 1895.—Patient, by her own wish, was discharged; there had been improvement.

Tab. 47 b.

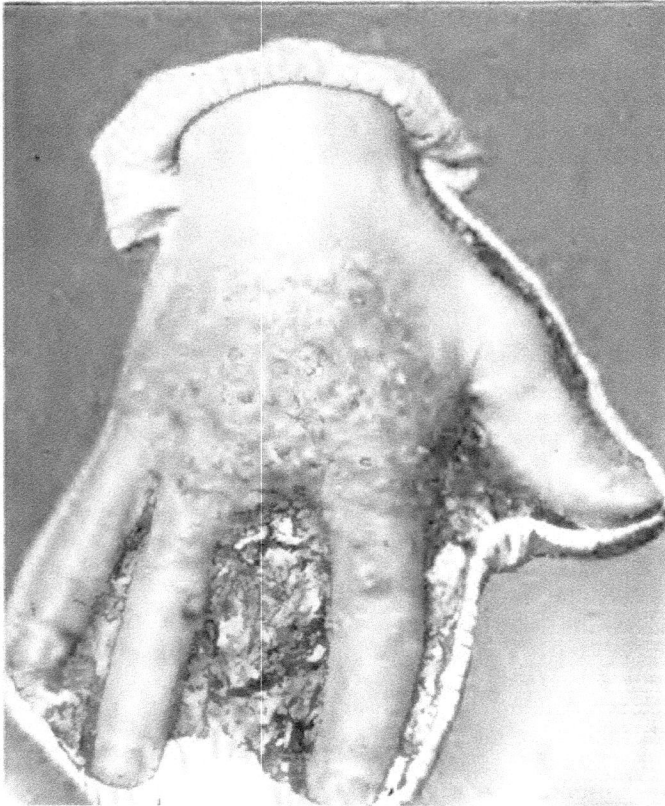

Chronic Tuberculosis of the Skin of the Leg (Lupus Tumidus).

P., aged 69 years, was admitted Dec. 25, 1894. On the skin of the lower leg were numerous warty papillomatous excrescences, furrowed, and with points of ulceration, out of which could be squeezed cheesy pus and blood. The development of the papillomatous growths was rapid, becoming quite extensive over this leg, there being also marked pigmentation. Two nodules were excised, examination of which disclosed the process to be a typical tubercle-deposit-formation in the granulation-tissue.

Treatment.—As the patient refused operative measures, injections of Koch's tuberculin were tried. Injections were made on Feb. 21 and 26, Mar. 3 and 14, and Apr. 4, each time one milligram. Reaction appeared after the first injection; temperature rising to 39° C., falling two hours later to 38° C.; a day afterward it had become normal. After the second injection the temperature rose to 40° C., and about the same elevation followed each of the succeeding injections. The patient always felt sick and weak for one or two days after each injection, but recovered rapidly. The local changes after the first injections consisted in increase of the swelling, congestion of the growths and their surroundings, and a melting away of the lesions. By the time of the last injection papillomatous growths and tumors were merely flat infiltrations; the local reaction showed itself by hyperemia of these areas.

The case is of interest for two reasons: First, the appearance of tuberculosis of the skin on an unusual site and the peculiar appearances and course of the growths; and secondly, the result of the treatment instituted.

Tab. 47 c.

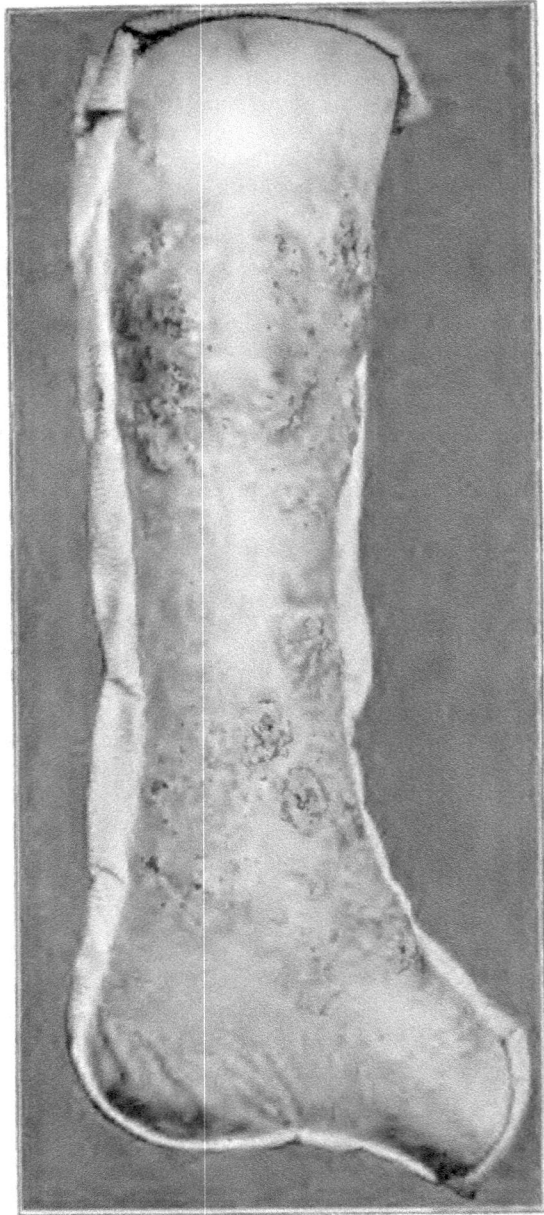

PLATES 48 and 48 a.

Lupus Vulgaris; Phlegmon.

J. A., aged 20 years, trunk-maker, admitted May 9. The patient has been sick since early childhood; the skin-disease is of about fourteen years' duration.

Status Præsens.—Patient is large, very anemic, and emaciated. Pulmonary tuberculosis; amyloidosis hepatis; nephritis. The left lower extremity is elephantiasic, thickened, and edematous; the dorsum pedis and the interdigital spaces covered with discrete and confluent ulcerations. Scattered groups of lupus-tubercles on the left thigh. On the mucous membrane of the cheek are several millet-seed-sized ulcers. On the right thigh, starting from a scattered group of lupus-tubercles, is a phlegmonous inflammation, with lymphangitis, which extends to Poupart's ligament.

May 15.—Severe pain in the left lower extremity; at the same time there was noted marked increase in the edema and the skin became rugous and wrinkled. A bluish-black discoloration of the toes developed, which rapidly spread. Death ensued in the night.

Autopsy.—In the biceps muscle of the thigh was an abscess the size of the double fist; lying between it and the bone was the femoral artery. Tuberculosis cutis (lupus verrucosus); lupus mucosæ oris; hypoplasia arteriarum; amyloidosis hepatis, lienis, et renum; nephritis subacuta. -

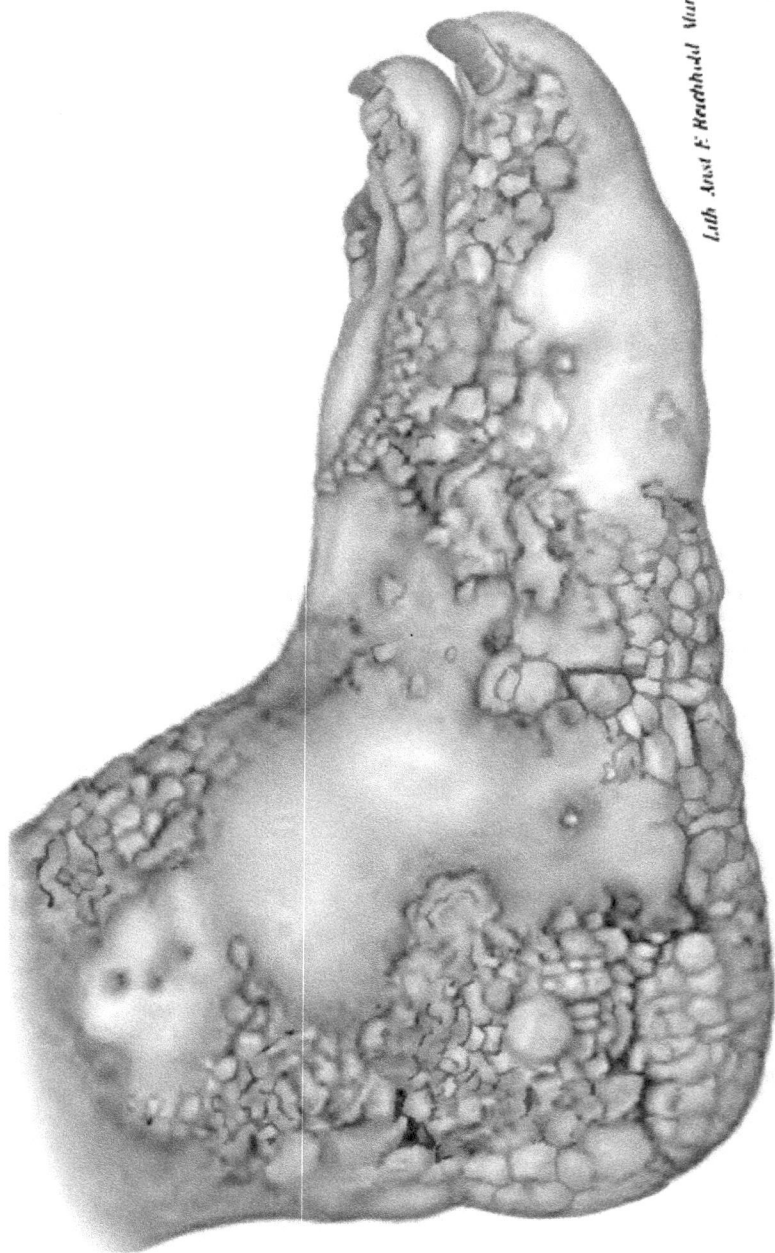

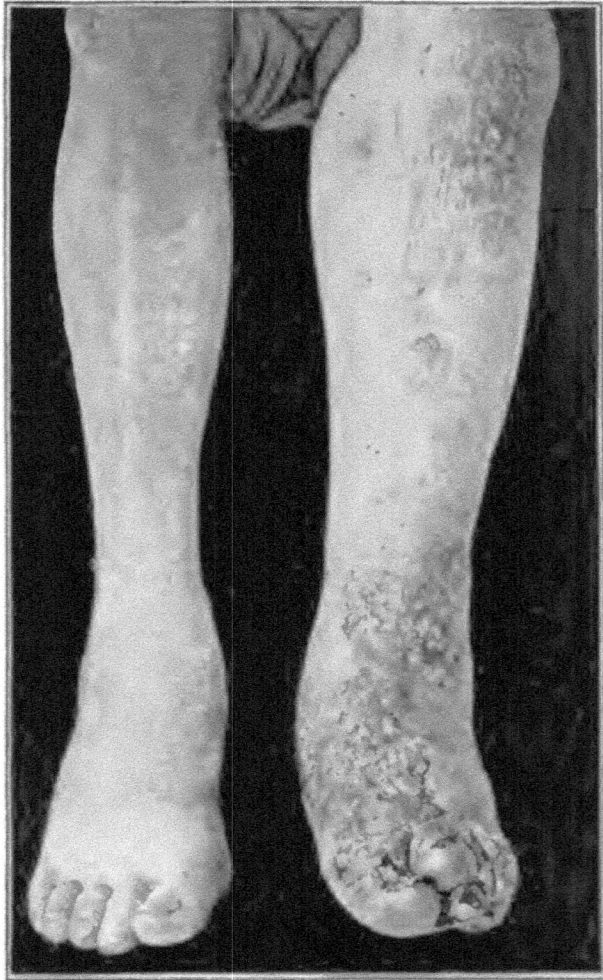

Tab. 48 a.

PLATE 48 b.

Lupus Vulgaris (Lupus Hypertrophicus).

M. C., aged 60 years, was received in the hospital in Sept., 1897. The patient was much debilitated and mentally depressed, and stated that for a year various parts of the face had been rapidly and consecutively attacked with considerable inflammatory swelling. He knew nothing of any earlier eruption, especially as he never had had any pain.

Status Præsens.—The face is deformed, the right eye almost closed, the cheeks and the nose, for the most part, the seat of elongated, furrowed scar-tissue; likewise the edematous upper lip. Between the eyebrows and root of the nose, over the left zygoma, over the right part of the left maxilla, and on the right cheek, are ulcerations not very much infiltrated, covered with crusts. The neighborhood of the right angle of the mouth and the lower lip are edematous and swollen; small points and areas of still greater thickening, in these regions, are recognizable by the touch. The mucous membrane of the upper lip and cheeks is much reddened, and here and there, near the edges, is eroded and even ulcerated.

The patient was not able to open the mouth and was artificially fed. His condition was somewhat improved after two weeks in the hospital, but he was then obliged to leave for home.

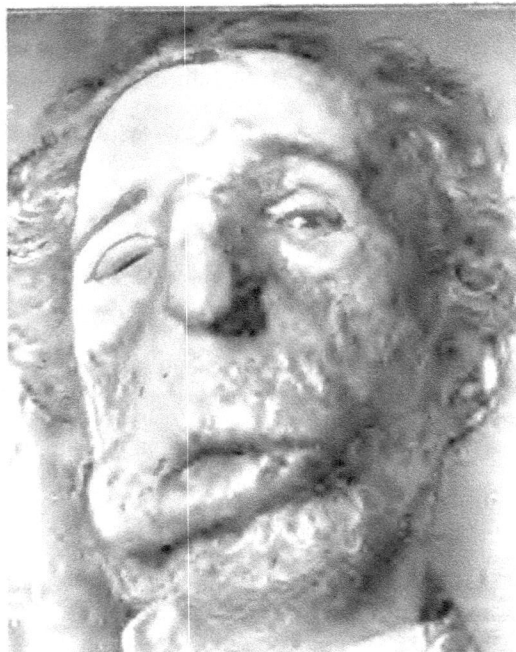

Tab. 48 b.

PLATE 49.

Tuberculosis Subacuta Mucosæ Oris.

K. J., aged 42 years, hotel-keeper, was admitted Feb. 8, 1897 The patient stated that for two years he has been sick. His trouble began with a swelling of the right half of the lower lip, which gradually spread superficially. At the same time there appeared ulcerations on the mucous membrane of the mouth. His disease was considered an actinomycosis, and the ulcers were cauterized, partly with the Paquelin cautery and partly with acid. There was slight improvement, which did not, however, continue, and the past month there has been a positive aggravation.

Status Præsens.—The patient is of medium size, well nourished, and strongly built. The left cheek is swollen, and on the inner side, to the extent of a silver quarter, are found hemp-seed- to small pea-sized papillary growths. The mucous membrane of the lips as well as that of the left cheek near the mouth-angle is swollen and the seat of numerous millet-seed- to hemp-seed-sized, and several larger, irregularly-shaped ulcers, covered with grayish-yellow adherent deposit. The gums of the upper and lower jaws show similar changes.

The lungs, except the apex of the right, are normal, over this latter the percussion-sound is shorter and duller; auscultation gives râles, whistling, and irregular inspiration and expiration.

The plaques in the mouth are very painful, and in them the presence of tubercle-bacilli was demonstrated.

Treatment.—Applications of a 1 per cent. sublimate solution, and cauterization with 20 per cent. lactic acid; both gave considerable pain. The disease progressed, showing no disposition toward improvement, and the patient, at his own request, was discharged after a stay of thirteen days.

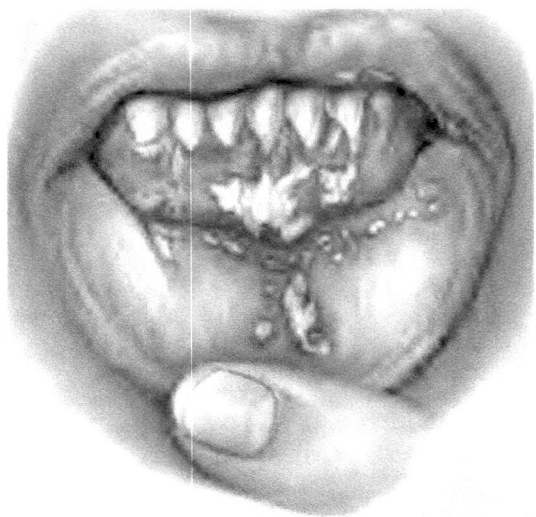

PLATE 50.

Panaritium Tuberculosum.

W. J., aged 48 years, with advanced pulmonary tuberculosis.
The patient was of strong bony structure, but cachectic. He
stated that in 1891 a small ulcer appeared at the nail of the
middle finger of the right hand, which since that time had persisted and gradually spread over the third and second phalanges.
The finger is thickened toward the end, especially at the joint.
The skin is livid. The nail is in process of being cast off, the
base being yellowish and to some extent broken off, and lifted
up from the matrix. The uncovered portion shows ulceration
covered with crusts. On several places are to be seen small
pea-sized to bean-sized ulcers covered with granulations; in
addition there are several crusted ulcers. The movability of
the finger, except between the first and second phalanges, is
compromised. The patient has boring- and tearing-pains in
the affected parts, at which times the finger always swells and
breaks out in one or two spots, from which pus exudes; this
takes place mostly about the nail. This pus-formation has,
he stated, only been noticeable the past several months; during
this period, too, the bones of the part have become involved.
Formerly the finger was dry, not so swollen, and less painful.

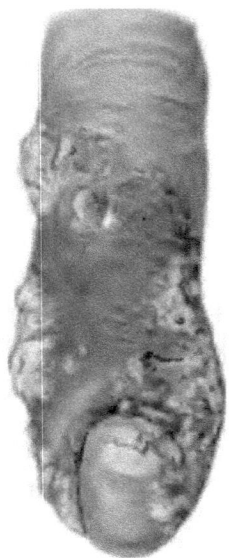

PLATE 51.

Tuberculosis Cutis.

F. B., aged 54 years, porter, was treated in the dispensary-department. Patient was otherwise healthy and strong, although at the time somewhat emaciated.

Status Præsens.—The disease occupies the region between the index and middle fingers of the left hand; the affected area is elevated, mildly inflammatory, and irregularly furrowed, and of the size of a silver half-dollar. To the touch the individual nodules are somewhat elastic, and by strong pressure sebaceous-looking material in small quantities exudes from the furrows. The growth is not painful. The disease began six months previously, and had, at first, the appearance of a wart. Patient was not aware of any cause for the disease.

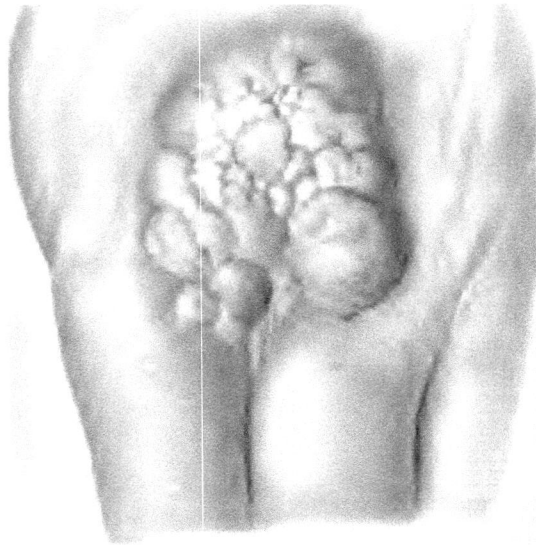

Tab. 51 a.

PLATE 51 a.

Lepra.

(FROM THE CLINIC OF PROFESSOR DE AMICIS, NAPLES.)

P. F., from Bisceglia (province of Bari), aged 43 years, baker, married, was admitted Jan. 12, 1895, and discharged Jan. 18.

History.—There was no hereditary tendency. In his home-region were several lepers. Two brothers had the same disease before he had it; neither had been outside of the country. Shortly before the disease appeared the patient had married; he has no children; his wife remains healthy. The disease first showed itself when he was twenty-seven years old, with the appearance of bullæ, at first on the legs and then on the upper extremities, with resulting sluggish ulcerations which did not heal. Gradually the face began to share in the process, becoming considerably disfigured.

Status Præsens.—Head-hair normal; everywhere else the hair has disappeared, even in the genital region. The skin-color of the face is slightly bluish in tinge, and the appearance considerably distorted and disfigured. Over the forehead, and especially the eyebrows and glabella, the skin exhibits numerous reddish-brown, more or less confluent, lumpy infiltrations, with numerous furrows. Over the malar bone there is capillary enlargement. The appearance of the nose is completely changed; the bridge is flat and sunken, especially at the junction of the bone with the cartilage, where a semicircular furrow is noticeable. The alæ nasi are somewhat infiltrated, and the nasal openings contracted or closed, especially the right one; the bony septum is absent. The lips are bloated. The chin shows infiltration, with many furrows, but less numerous and less deeply than the forehead.

There are many infiltrations and yellowish spots on the upper extremities, on which ulcerations, torpid in character, are also observed, more especially on the extensor surfaces of the right forearm and on the dorsal surface of the right hand. There are likewise similar conditions on the elbow and dorsal surface of the left hand. On the buttocks and lower extremities the same lesions are to be observed. The scrotum, the skin of the penis, the prepuce, and the glans penis are infiltrated. The mucous membrane of the hard and soft palates, the tongue, epiglottis, and the arytenoid folds show more or less gray diffused infiltration. Everywhere anesthesia.

Bacteriologic examination of blood from the infiltrations showed the Hansen bacilli.

PLATE 51 b.

Lepra.

(FROM THE CLINIC OF PROFESSOR DE AMICIS, NAPLES.)

D. E. F., of Ischitella (province of Foggia).

History.—The patient came of a fisherman's family. His parents are living and healthy; his grandfather suffered from the same disease. Of the seven brothers and sisters, the patient was the third. His four brothers, the second, fourth, sixth, and seventh children, have also the same disease; the remaining two are healthy. The affection first appeared in 1880, when he was twenty years of age; he came under medical observation in May, 1892.

The disease began with attacks of chilliness, with elevation of temperature following, and the appearance of red spots, at first on the upper extremities, with subsequent tubercle-infiltration. After some months it showed itself in the face, and later also on the lower extremities; in this latter region accompanied, for a time, with severe muscular pain. Infiltration followed soon after the macular lesions or stage, especially in the face, of which a considerable part is involved and much disfigured. On the trunk the disease spread in the form of spots and tubercles. Recently the voice has been hoarse and weak.

Status Præsens.—An examination shows that there are numerous tubercle-infiltrations, scattered and confluent, separated by more or less deep furrows, anesthetic, reddish, and brick-red in color, covering the forehead, the alopecic eyebrows, cheeks, lips, and chin, so that the face has lost its original appearance and is now suggestive of leontiasis. Also on the upper extremities are to be seen similar changes, tubercles and spots, more or less confluent, especially on the elbows and dorsal surfaces of the hands and fingers, where also some sluggish ulcerations are to be observed. On the trunk and lower extremities similar lesions are also to be seen, but are less abundant and less confluent.

The mucous membranes of the mouth and of the hard and soft palates are covered with grayish papular infiltration; likewise the epiglottis and vocal cords. There is everywhere anesthesia.

Bacteriologic examination of blood from the infiltrations gave the Hansen bacilli.

PLATE 51 c.

Lepra.

(FROM THE CLINIC OF PROFESSOR DE AMICIS, NAPLES.)

D. S., from Marsala, lumberman, unmarried.

History.—Patient is of robust constitution and of good growth and size. His parents and the relatives of the family are healthy. At his home some cases of the disease are observed. The patient had never travelled. Toward the end of May, 1878, in his nineteenth year, he noticed the gradual appearance of scattered spots, of varying size, of reddish-brown color; at first on the lower extremities and then on the upper extremities. Soon afterward these were followed by similarly-colored milletseed- to lentil-sized papular elevations, which from the beginning were accompanied by a burning sensation, and later by itching. Finally on the affected parts the sensibility was lessened. After one and a half years he observed falling out of his eyebrows and eyelashes. In Nov., 1880, the disease having been gradually progressing, he appeared for the first time at the clinic; then again in 1881, 1886, and 1888, where he at the last time presented the following lesions and symptoms.

Status Præsens.—*Head.*—Striking tubercle- and nodular infiltration of the forehead, the region of the eyebrows, glabella, the nose, the lips, cheeks, and chin, of the size of a silver quarter or smaller; many superficially ulcerated and covered with blood-crusts. Complete loss of hair of the face. An almond-sized nodule on the left upper lid and several smaller, millet-seed-sized, on the right lid; an almost completely pedunculated nodule on the conjunctiva bulbi of the left eye, with accompanying keratitis; the same on the right eye. Nodular infiltration of the ear-muscles. The hairy scalp is normal.

Trunk.—On the breast disseminated reddish-brown spots with small papular elevations, especially in the neighborhood of the nipples. Some larger, diffused, brick-red spots with symmetric papulotubercular elevations on the shoulders and loins.

Neck.—Two nutmeg-sized, half-rounded nodes with smooth surface, one on the left side on the upper third of the sterno-cleido-mastoideus muscle, the other on the right side over the mastoid process. Several smaller nodes on the nape of the neck.

Upper Extremities.—Almost the entire surface is reddish-brown, with the exception of some disseminated places where

Tab. 51 b.

the skin is normal. From the shoulder-joint down to the elbow-groove can be seen, on the extensor surface, nodular, lupus-like elevations; these are on the upper part confluent, and on the left side give rise to nut-sized ulcerated nodes. On the forearms are numerous nodules, nodes, and tubercles, some somewhat pedunculated, smooth, and with shining surface; others closely ribbed and covered with bran-like scales. They are irregularly distributed, with the greater number upon the extensor side. In the wrist-region are large confluent tubercles bunched closely together; traversed with furrows and scars and beset with ulcerating nodules. Owing to a deep infiltration of the dorsal aspect the hands are much maimed, and the fourth and fifth fingers are held in a flexed position. Atrophy of the interosseous, thenar, and antithenar muscles.

Lower Extremities.—As on the upper extremities, the eruption is here widespread, extending from the gluteal region to the feet. In addition to numerous spots, numerous elevations and tubercles are to be seen. Some are pedunculated, others ulcerated and crateriform; the largest being over the tendon Achilles, along the anterior border of the tibia, and on the front part of the knee. The skin on the posterior surface of the legs is irregularly infiltrated. From the dorsal surface of the feet to the ends of the toes are exceedingly numerous elevations and tubercles, crowded closely together: on the right foot is an isolated giant tubercle the size of a silver quarter. Also on the soles are bunched tubercles.

Genitalia.—Tubercles on the skin of the penis and the scrotum. Extensive, almost almond-sized infiltrations in the right epididymis; in the left epididymis still larger nodes and numerous smaller tubercles on the surface of the testicles.

Mucous Membranes.—The entire hard palate is covered with an ulcerated, grayish covered granuloma. There are also ulcerations and scars on the soft palate and tonsils. The uvula and nasal mucous membrane, especially on the septum, are the seat of extensive infiltrations.

Lymphatic System.—The glands in the neck and in the groin are markedly enlarged, the latter almost as large as a fist.

Sensibility.—Sensations of touch, heat, and pain very much diminished and in some places entirely wanting.

Urine.—The urine is ropy and rich in mucin.

Microscopic examination of blood from the diseased nodes disclosed numerous lepra-bacilli.

Tab. 51 c.

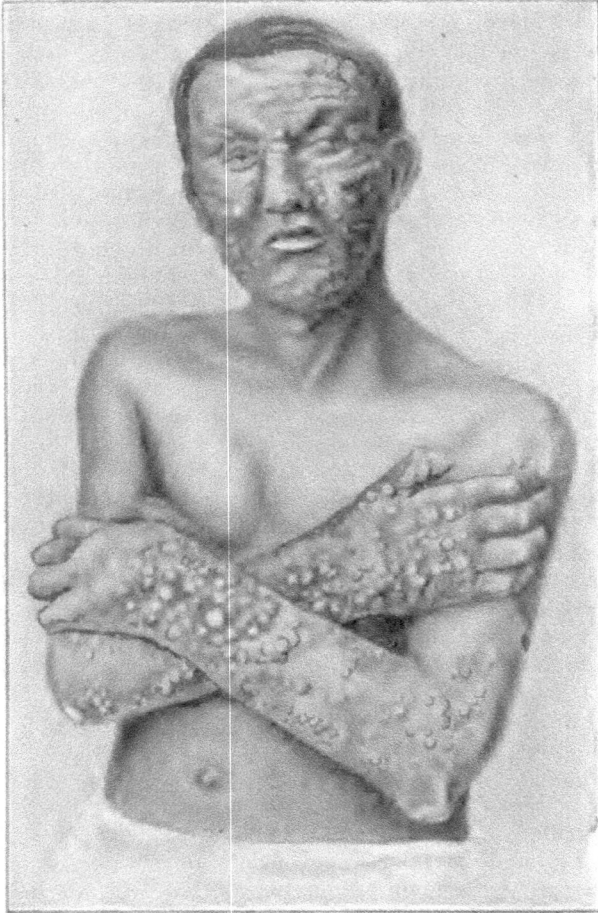

PLATE 52.

Carcinoma Lenticulare.

S. A., aged 74 years, admitted July 6, 1896, stated that one year previously the left breast began to harden.

Status Præsens.—The head is directed toward the left, and it can be turned only to a moderate degree, and that with difficulty. The skin of the left breast, of the neck-region, and extending to the back and to the face, is the seat of a yellowish-red to violet-colored, tough, hard, in part cicatricial-looking growth or tumor. The border is, especially at the lower part, sharply defined against the healthy skin and slightly elevated. Toward the face and back its junction with the normal skin is not so clearly recognizable. The left side of the face is edematous. The submaxillary, supraclavicular, and infraclavicular glands are hard, infiltrated, and enlarged. The opening of the mouth is somewhat hindered, owing to lack of complete movability of the lower jaw. Swallowing is likewise less easy than normally.

In the following two months no material change ensued. The face and shoulder varied somewhat as to the amount of edema.

On Sept. 2, two months after admission, the patient died with symptoms of collapse.

Autopsy.—Diffused and lenticular sarcoma of the skin over the left breast, arising from a carcinomatous mammary gland; sarcomatosis of the pleuræ, peritoneum, and uterus.

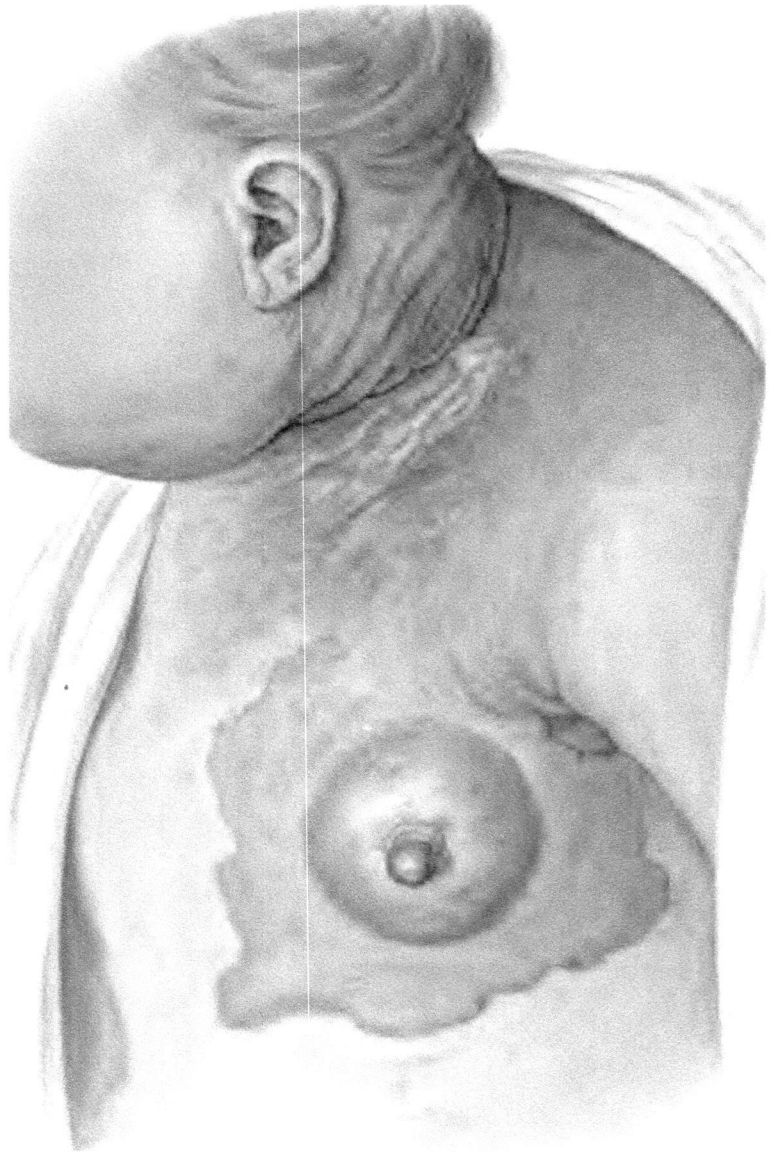

PLATE 53.

Epithelioma.

W. M., aged 60 years, cook, came under treatment May 5, 1897. The patient first noticed the disease about five months previously. It had given rise to no pain. She had always enjoyed good health, except having, when aged 40, a peritonitis, from which she made a good recovery. She has given birth to one child. Menstruation ceased five years ago

Status Præsens.—The patient is of moderately strong build and fairly nourished. On the lower part of the left labia majora is a dollar-sized ulceration with an infiltrated and elevated base. The surface is irregular, red, and uneven, with here and there whitish spots. The secretion is scanty. There is no enlargement of the inguinal glands. The opposite lip is not involved. Above the growth, on the same side, toward the vagina, is a bean-sized nodule with epithelial proliferation and beginning central destruction.

Treatment.—Under chloroform the diseased area was excised and the patient was discharged cured on June 5.

In May, 1898, about a year later, there was a recurrence with involvement of the inguinal glands. Another operation followed, healing taking place in six weeks.   .

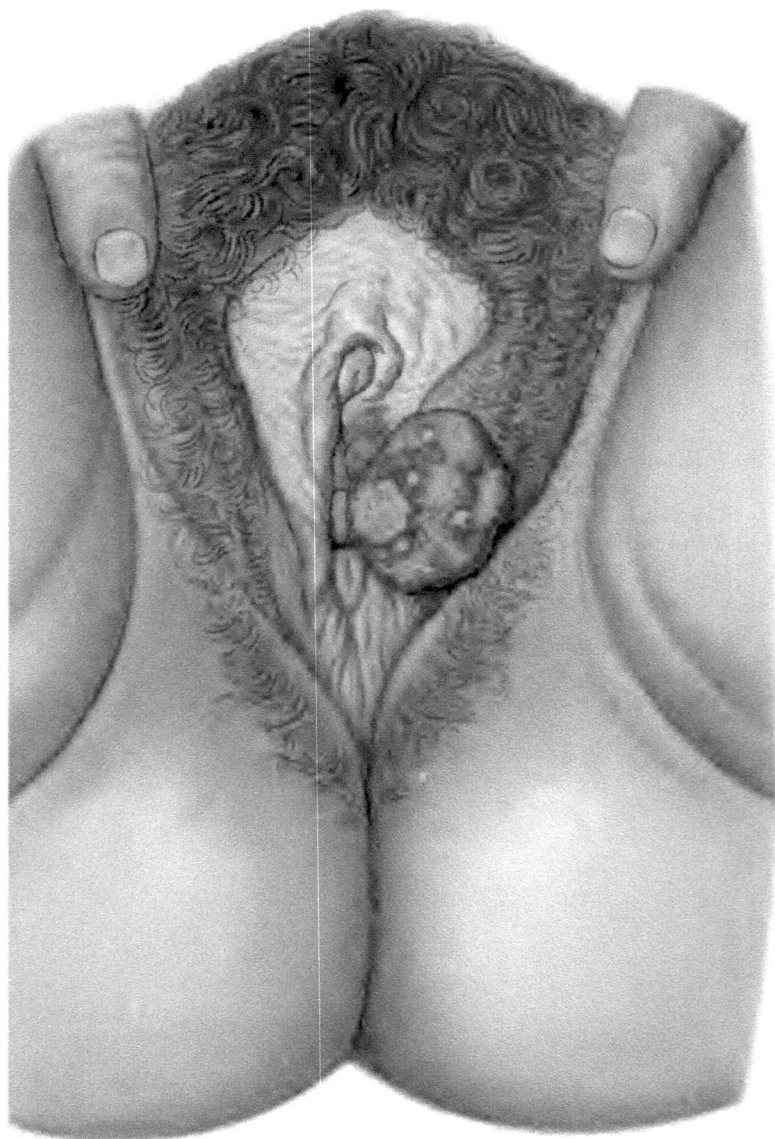

PLATE 54.

Carcinoma Penis.

(CASE FROM PROF. ALBERT'S CLINIC.)

N. N., aged 51 years, admitted July 9, 1890.

History.—Fifteen years previously patient met with an accident, suffering an injury to scrotum and penis. The wound healed; subsequently a growth began between the scrotum and base of the penis. Two years ago an ulcer appeared on the penis, which gradually enlarged and gave rise to considerable pain. Lately the patient has lost a good deal of flesh.

Status Præsens.—The penis is hard, misshapen, and the seat of fissures and ulcers; is a little less than five inches long, and four inches in circumference. On lifting the organ a palm-sized ulcer is seen, with hard borders and covered with irregular sluggish granulations. The inguinal glands of both sides are enlarged.

Treatment.—Amputation of penis; removal of inguinal glands. The urethra was dissected out and stitched to the perineum.

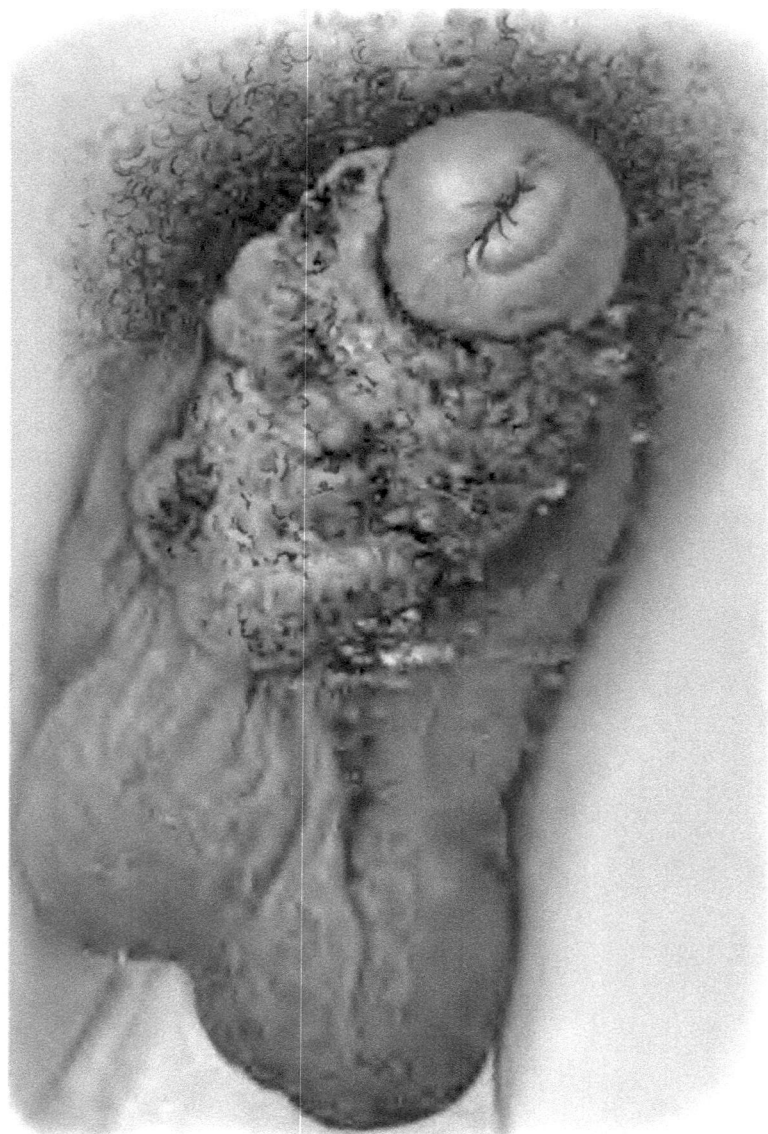

PLATE 55.

Carcinoma Penis.

(FROM PROF. ALBERT'S CLINIC.)

According to the statement of the patient, the disease had existed six months.

Status Præsens.—The patient is strongly built, but emaciated. The skin of the penis is covered with scars, partly pigmented and partly changed into thick, tough infiltration; a high degree of phimosis exists. On the under half the infiltration is continuous and the base irregularly excavated and ulcerated, and the whole mass is hard and dense. The lymphatics on the dorsal side of the penis, and the inguinal glands, are swollen and hard.

Treatment.—Partial amputation of the penis, with plastic operation for urethra.

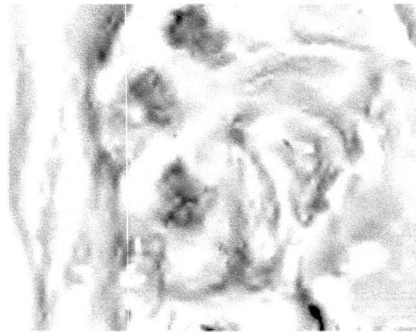

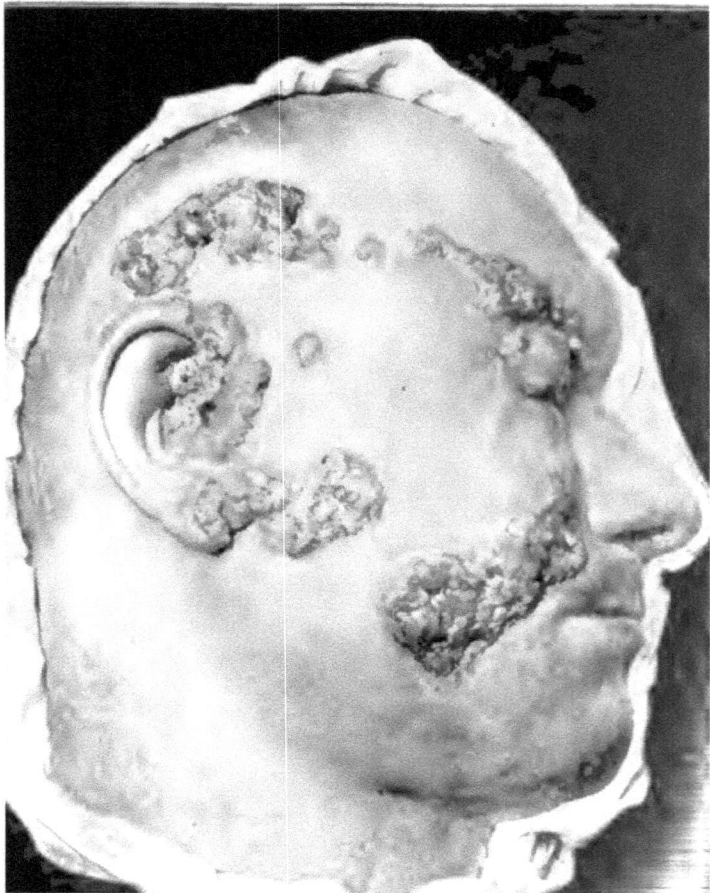

PLATE 55 a.

Epithelioma Cicatrisans.

J. J., aged 55 years, day-laborer, admitted Sept. 28, 1892. The disease began six years previously, on the right temple, as a small nodule, from which point it spread as a continuous ulcer on to the cheek and to the right eyelid.

Status Præsens.—The right cheek is, from the ear-muscle posteriorly to the nasolabial fold anteriorly, upward toward the attachment of the masseter muscle, and downward to the inferior maxilla, changed into a smooth whitish scar. The border of this area consists of an almost continuous ulcer, somewhat elevated, with a base showing but slight infiltration. The base seems made up of anemic granulation-tissue. The disease has eaten through the upper eyelid, and the lid is somewhat drawn outward by the scar-tissue. The patient complained of stinging-pain occasionally in the ulcerated part.

At intervals the proliferation was curetted, and in this way for a time destructive action or progress was stayed. On Dec. 15 the growth was investigated histologically and the diagnosis of epithelial carcinoma confirmed.

Under anesthesia the ulcerated surface was thoroughly curetted and then cauterized with the Paquelin cautery, so that all, with the exception of a linear ulcer at the corner of the mouth, healed and scarred over.

Scarcely four weeks had elapsed after this operation before the remainder of the eyelid at the inner canthus broke down; a new destructive action was also observed over the zygoma, and the epithelial masses at the corner of the mouth began to grow considerably. The patient's weight, with slight fluctuation, remained at 54 kg. The ulcerated surfaces extended and involved the scar-tissue. The eyeball was attacked, and lay in the orbital cavity surrounded by epithelial necrotic masses. The patient complained of increasing pain, which could only be relieved by constant use of morphin.

Dec. 22.—In the center of the extended ulcers small islets of scar-tissue are again to be seen, although the disease has now spread over the entire chin and also over the middle of the nose.

The patient was finally obliged to return to his home, and

left the hospital on Sept. 21, 1894. The case was under observation for two years.

The case is remarkable in that in the entire eight years of its existence there had been no tendency to change its character. Further to be noted were the slow course and the tendency to cicatricial formation in the central parts. Later, however, not only did the ulcerated parts advance, but the already formed scar-tissue again gave away. This disposition to cicatricial formation was shown again and again, but the disease slowly progressed. The patient, from constant pain, became more and more emaciated.

PLATE 56.

Tinea Favosa.

S. L., **aged** 25 years, admitted Aug. 18, 1896. For a number of years scalp-eruption and hair-loss had existed.

Status Præsens.—The scalp-hair, with the exception of a narrow fringe posteriorly, has entirely disappeared. The scalp-skin is covered in many places with sulphur-yellow, kidney-shaped crusts. Between these larger crusted areas are scattered pinhead- to small pea-sized straw-yellow lesions; the same also on the shoulder.

After three months' treatment the scalp is clean, and no new lesions or crusts have appeared. It remained in same condition when discharged Jan. 5, 1897.

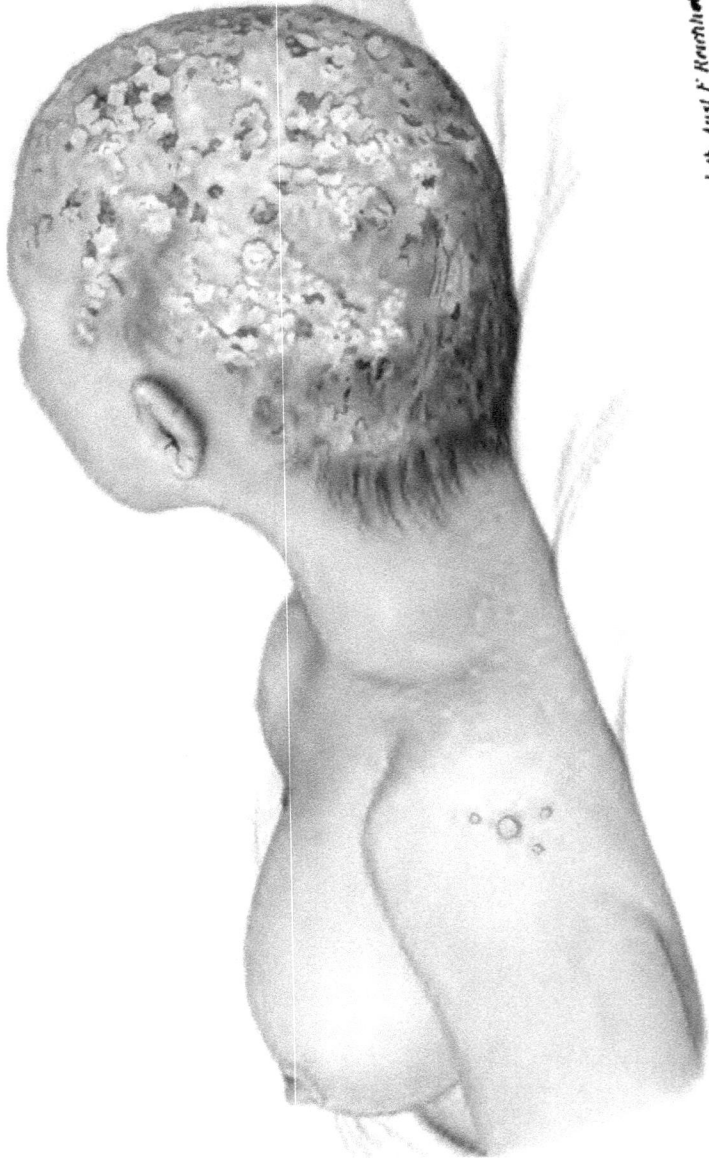

PLATE 57.

Pityriasis Maculata et Circinata.

S. F., aged 18 years, admitted Feb. 13, 1896. One day before admission the patient noticed that the spots had appeared. His attention was first called to them by the itching.

Status Præsens.—The thorax, abdomen, and the flexor surfaces of the extremities are the seat of numerous efflorescences. On the lower belly and pubic region the spots are pale red, the larger of which show a central whitish epidermic scale. The larger number have already paled in the central portions, showing peripherally faintly-wrinkled epidermis, and here and there partly-detached scales. The border of the patches is slightly elevated, the epidermis of which is smooth and reddened. Similar features are presented by the patches on other parts.

[In the German edition this plate is described under the heading of "herpes tonsurans maculosus et squamosus," a variety of ringworm. American and English observers, however. consider the disease as here pictured as pityriasis maculata et circinata, a disease entirely independent of the ringworm-fungus.—Ed.]

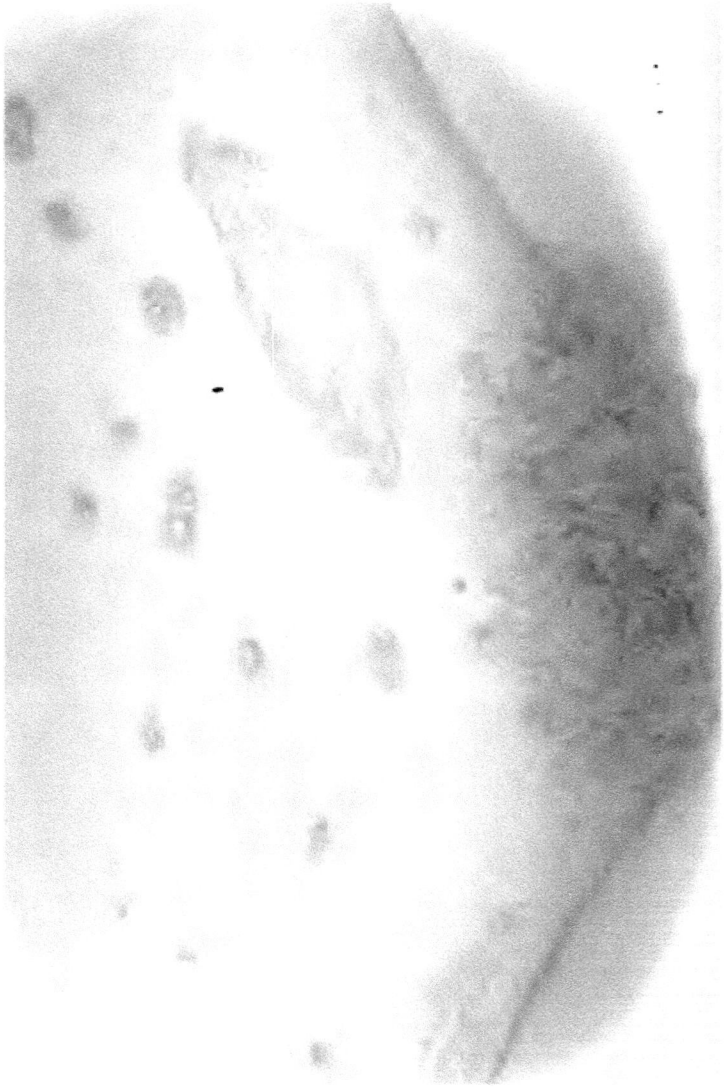

PLATE 58.

Tinea Trichophytina Corporis (Tinea Circinata).

L. W. admitted Nov. 16, 1895; discharged cured Nov. 23. Eight days previously patient noticed the central part of patch; since that time it had gradually enlarged to its present dimensions.

Status Præsens.—Upon examination is found on the right breast a half-dollar-sized efflorescence, the center of which is yellowish-red and slightly scaly. The peripheral part of the patch is somewhat elevated, slightly crusted and scaly, and reddish in color. The outermost edge is sharply defined against the sound skin and is of a bright-red hue. There is itching, but not to a troublesome degree.

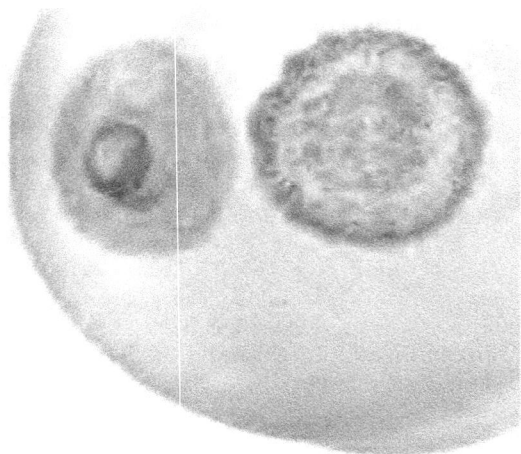

PLATE 59.

Tinea Trichophytina Corporis (Tinea Circinata).

S. F., aged 18 years, locksmith. Under observation from May 19 to 28. **Fourteen** days previously the disease had appeared on the face, **and** during the past week on the left upper extremity. Horses **were kept** in the **house** in which patient lived.

Status Præsens.—On the **face, and** especially **on the** left side, **are** numerous pustular efflorescences, varying **in** size from a pinhead to a **pea,** many covered with **a** yellow-brown **crust.** On the flexor side of the **left** forearm, close to **the** hand, is a large, rounded, infiltrated, **reddish** patch with an elevated periphery; **inside the border are a** number of papules and vesicles. **Cultures** were made with **the** contents of the vesicles and the trichophyton demonstrated.

Treatment.—Lassar's paste for the **face;** naphthol-sulphur paste with resorcin **for** the patch **on forearm.** Complete cure in eight days.

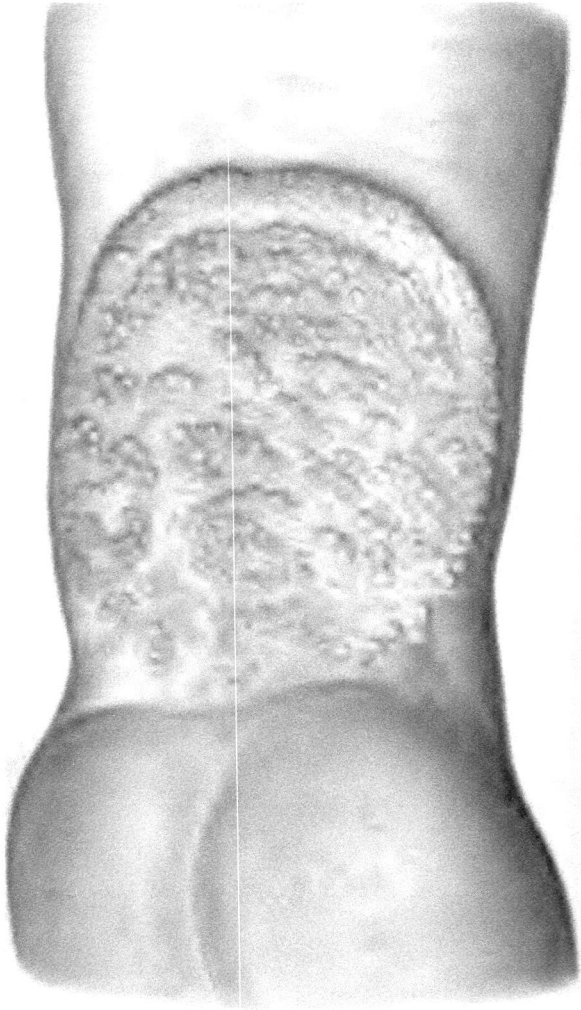

PLATE 60.

Tinea Versicolor.

J. N., aged 20 years, workwoman, admitted Aug. 18, 1897.

Status Præsens.—Over the breast are to be seen numerous, variously-shaped and -sized, yellowish-brown patches. Slight branny scaliness is observable in some, and the epidermal covering is readily scratched off. The color is pale yellowish-brown to a darker brown—the darker color being more pronounced at the edges.

Treatment.—Naphthol-sulphur soap, sapo viridis, and dusting-powder of rice-flour. Cure.

Tab.60.

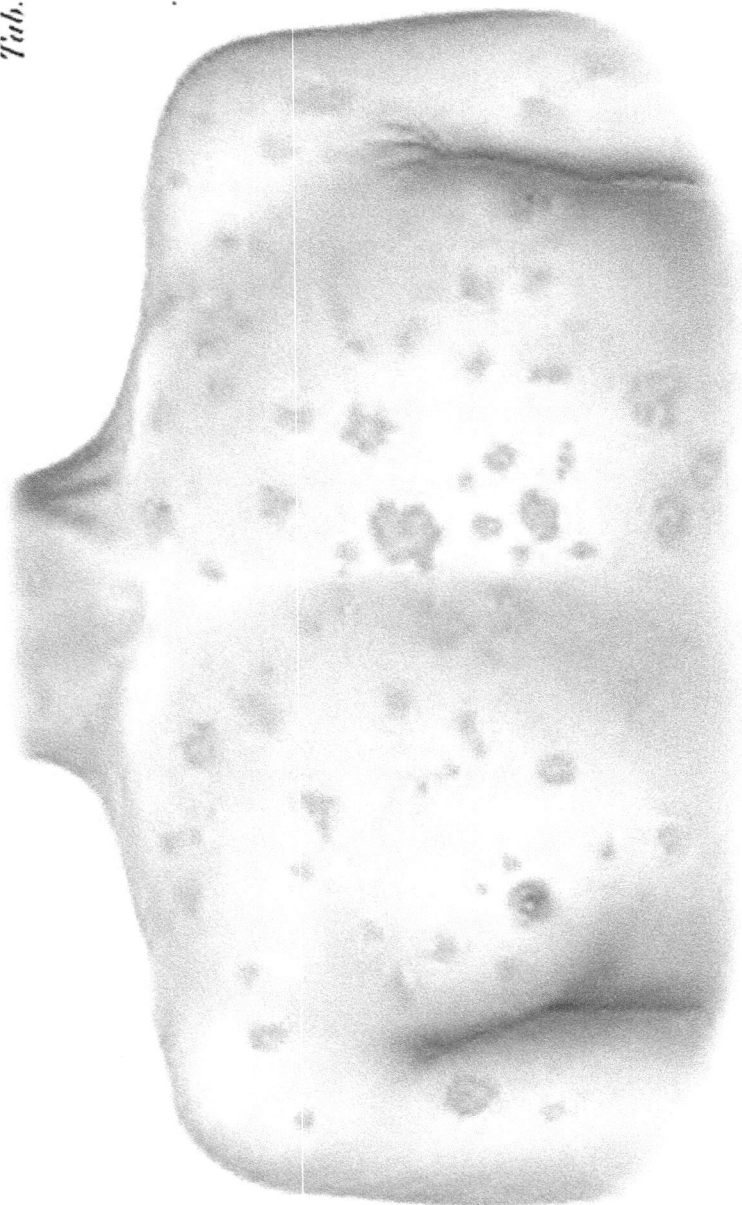

PLATE 61.

Actinomycosis.

D. A., aged 42 years. The patient was in the hospital in Aug., 1892, but returned to his home. As his condition had gradually grown worse, he was on request again admitted on Oct. 28, 1892.

Status Præsens.—The patient is very pale, emaciated, and complains of difficulty in breathing and swallowing. Lungs and heart normal. The entire left side of the neck, from the lower jaw down over the supraclavicular fossa, is made up of numerous elevations and depressions. The whole area is hard. Between these depressions the skin is infiltrated and correspondingly raised, and the seat of numerous fistulous tracts of varying depths; out of which there oozes thick pus containing the fungus, appearing as minute grayish-white or yellowish granules. The skin over the lower part of this infiltrated area is of a dirty violet-gray color. Immediately over the left collar-bone is a nut-sized fluctuating tumor covered with pale violet-colored skin; also on the right side of the neck, in the supraclavicular fossa, is a similar growth.

Treatment.—Patient was treated by incisions and the injection of ½ per cent. corrosive-sublimate solutions, and the parts kept covered with antiseptic bandages. Some improvement ensued. The patient insisted upon leaving the hospital three weeks after admission.

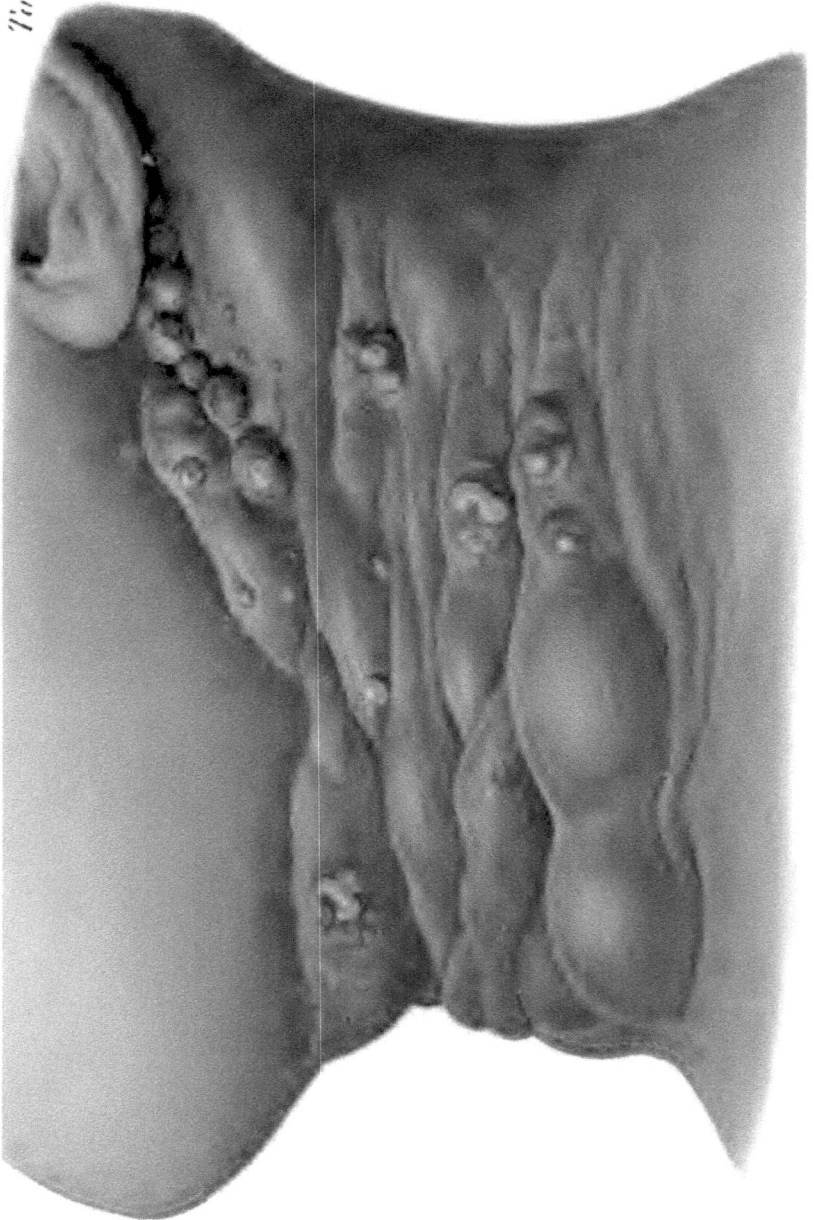

Tab 9

PLATES 62 and 62 a.

Scabies.

H. M., female, admitted **Aug. 12, 1897.** The patient stated that itching, especially **at night, became** noticeable six weeks **previously,** although **most of the pustular** lesions had appeared **more recently.**

Status Præsens.—The whole surface is the **seat of** irregularly-scattered scratch-marks and excoriations; **and** the **extremities are** covered with numerous discrete pustules, mostly crusted. **The dorsal surface of** both hands is studded with **well-filled pustules and** pus-containing blebs; **in some** places, more particularly on the fingers, **they have been scratched** away and given place to raw-looking abrasions.

Treatment.—Wilkinson's ointment; cure.

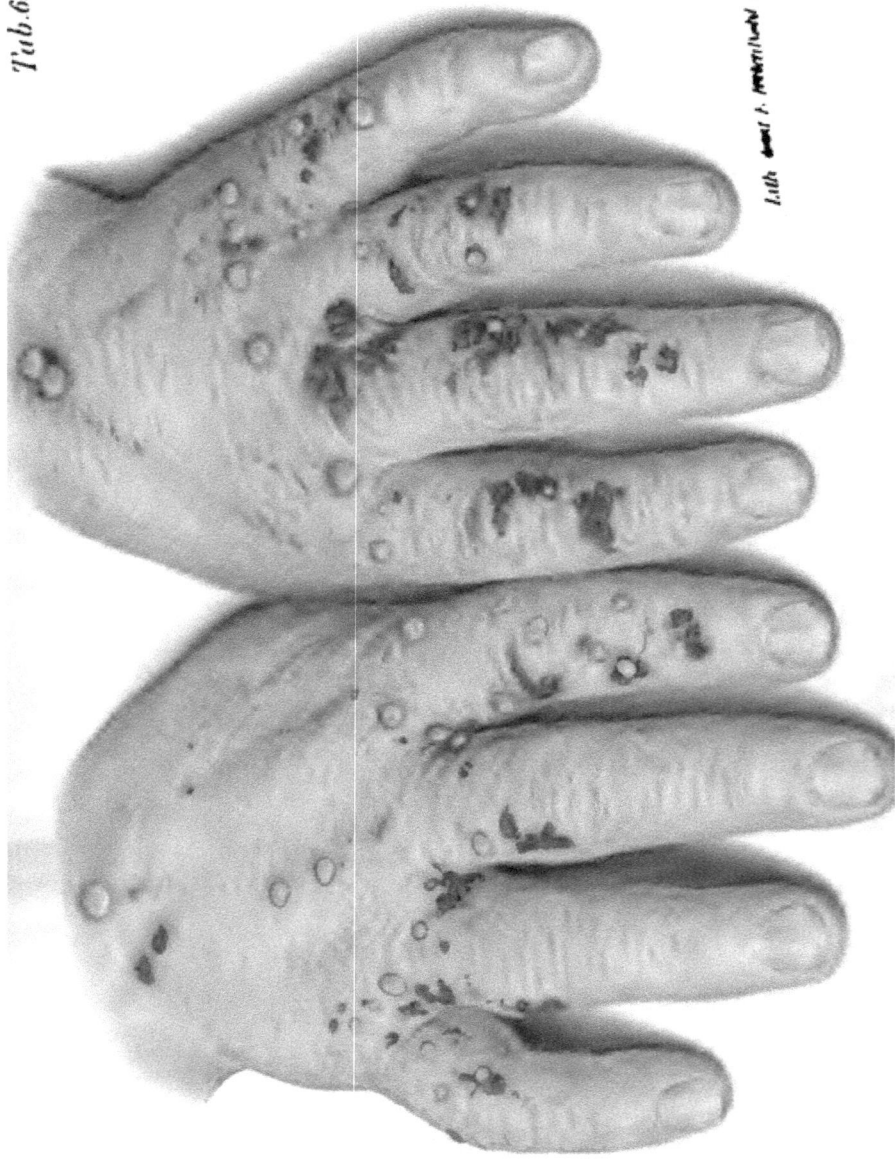

Tab. 62.

Lith. Anst. f. Mnnmuun, Münrhen

Tab. 62 a.

PLATE 63.

Maculæ Cæruleæ; Phthiriasis.

O. F., aged 33 years, baker's helper, admitted Aug. 21, 1897.

In the pubic and axillary regions numerous Pediculi pubis (crab-lice) are present, and ova may be observed attached to the hair-shafts. In addition, in the inguinal region, from the effects of scratching and from applications of mercurial ointment, are to be seen minute excoriations. The body is covered with bluish, rounded and linear, elongated spots up to the size of a pea; the overlying epiderm is uninjured.

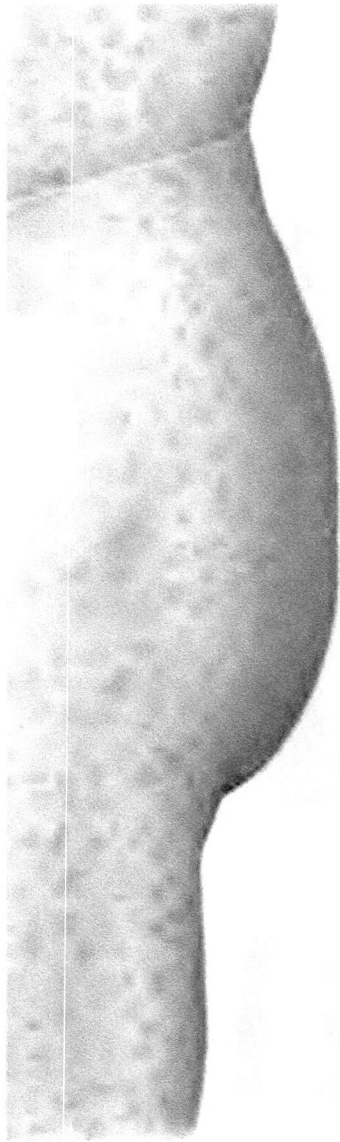

PLATE 64.

a. Nits (louse-eggs, ova), attached to the hair-shaft.

b. Head-louse.

c. Body-louse, clothing-louse.

d. Crab-louse.

e. A burrow (cuniculus).

f. Itch-mite egg.

g. Itch-mite, from beneath.

h. Itch-mite, from above.

i. Hair-follicle mite (Acarus folliculorum).

Tab. 64

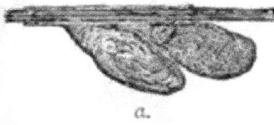

a.

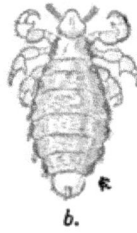

b.

c.

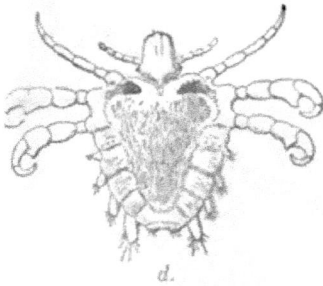

d.

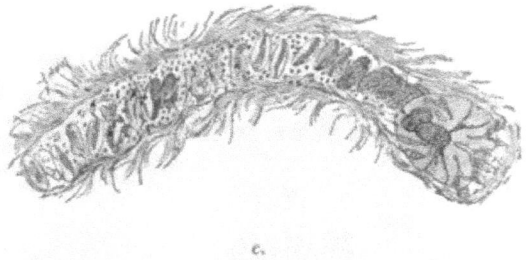

e.

f.

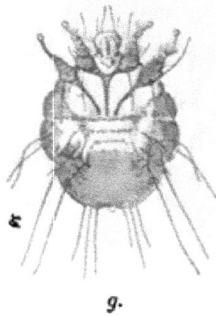

g.

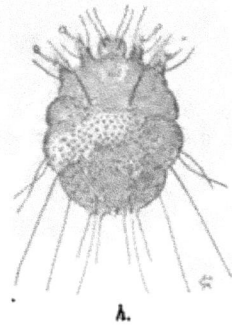

h.

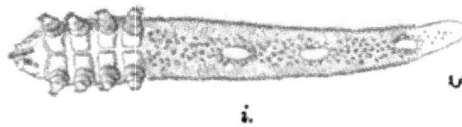

i.

PLATE 65.

a. Ray-fungus.

b. Molluscum epitheliale corpuscle; "molluscum body."

c. Trichophyton (ringworm-fungus) in scalp, hair- and outer root-sheath.

d. Microscopic picture of a hair in trichorrhexis nodosa.

e. Achorion Schönleinii (favus-fungus), from a favus-crust.

f. Microsporon furfur (tinea-versicolor fungus).

Tab. 65.

a.

b.

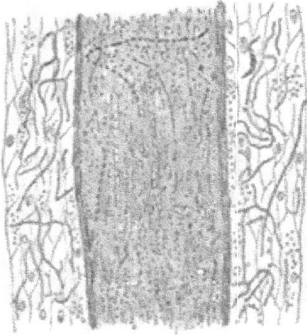

c.

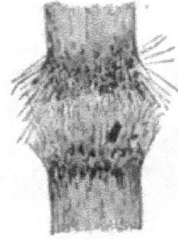

d.

e.

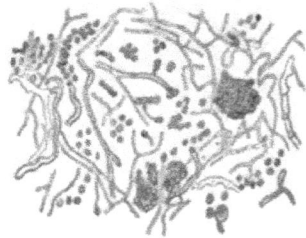

f.

PLATE 64.

a. Nits (louse-eggs, ova), attached to the hair-shaft.

b. Head-louse.

c. Body-louse, clothing-louse.

d. Crab-louse.

e. A burrow (cuniculus).

f. Itch-mite egg.

g. Itch-mite, from beneath.

h. Itch-mite, from above.

i. Hair-follicle mite (Acarus folliculorum).

Tab. 64

a.

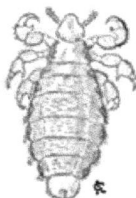

b.

c.

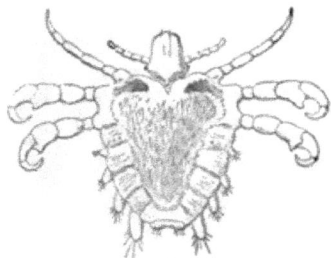

d.

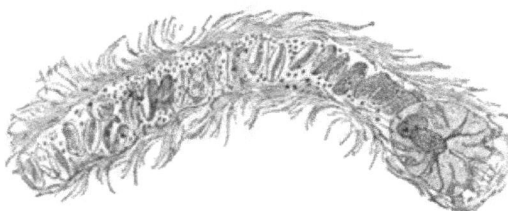

e.

f.

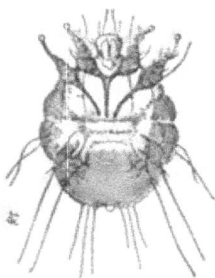

g.

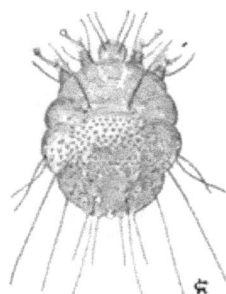

h.

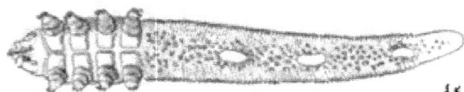

i.

INDEX.